Sensor and Data Fusion Concepts and Applications

Second Edition

Tutorial Texts Series

- *Sensor and Data Fusion Concepts and Applications, Second Edition*, Lawrence A. Klein, Vol. TT35
- *Practical Applications of Infrared Thermal Sensing and Imaging Equipment, Second Edition*, Herbert Kaplan, Vol. TT34
- *Fundamentals of Machine Vision*, Harley R. Myler, Vol. TT33
- *Design and Mounting of Prisms and Small Mirrors in Optical Instruments*, Paul R. Yoder, Jr., Vol. TT32
- *Basic Electro-Optics for Electrical Engineers*, Glenn D. Boreman, Vol. TT31
- *Optical Engineering Fundamentals*, Bruce H. Walker, Vol. TT30
- *Introduction to Radiometry*, William L. Wolfe, Vol. TT29
- *Lithography Process Control*, Harry J. Levinson, Vol. TT28
- *An Introduction to Interpretation of Graphic Images*, Sergey Ablameyko, Vol. TT27
- *Thermal Infrared Characterization of Ground Targets and Backgrounds*, Pieter A. Jacobs, Vol. TT26
- *Introduction to Imaging Spectrometers*, William L. Wolfe, Vol. TT25
- *Introduction to Infrared System Design*, William L. Wolfe, Vol. TT24
- *Introduction to Computer-based Imaging Systems*, Divyendu Sinha, Edward R. Dougherty, Vol. TT23
- *Optical Communication Receiver Design*, Stephen B. Alexander, Vol. TT22
- *Mounting Lenses in Optical Instruments*, Paul R. Yoder, Jr., Vol. TT21
- *Optical Design Fundamentals for Infrared Systems*, Max J. Riedl, Vol. TT20
- *An Introduction to Real-Time Imaging*, Edward R. Dougherty, Phillip A. Laplante, Vol. TT19
- *Introduction to Wavefront Sensors*, Joseph M. Geary, Vol. TT18
- *Integration of Lasers and Fiber Optics into Robotic Systems*, Janusz A. Marszalec, Elzbieta A. Marszalec, Vol. TT17
- *An Introduction to Nonlinear Image Processing*, Edward R. Dougherty, Jaakko Astola, Vol. TT16
- *Introduction to Optical Testing*, Joseph M. Geary, Vol. TT15
- *Sensor and Data Fusion Concepts and Applications*, Lawrence A. Klein, Vol. TT14
- *Practical Applications of Infrared Thermal Sensing and Imaging Equipment*, Herbert Kaplan, Vol. TT13
- *Image Formation in Low-Voltage Scanning Electron Microscopy*, L. Reimer, Vol. TT12
- *Diazonaphthoquinone-based Resists*, Ralph Dammel, Vol. TT11
- *Infrared Window and Dome Materials*, Daniel C. Harris, Vol. TT10
- *An Introduction to Morphological Image Processing*, Edward R. Dougherty, Vol. TT9
- *An Introduction to Optics in Computers*, Henri H. Arsenault, Yunlong Sheng, Vol. TT8
- *Digital Image Compression Techniques*, Majid Rabbani, Paul W. Jones, Vol. TT7
- *Aberration Theory Made Simple*, Virendra N. Mahajan, Vol. TT6
- *Single-Frequency Semiconductor Lasers*, Jens Buus, Vol. TT5
- *An Introduction to Biological and Artificial Neural Networks for Pattern Recognition*, Steven K. Rogers, Matthew Kabrisky, Vol. TT4
- *Laser Beam Propagation in the Atmosphere*, Hugo Weichel, Vol. TT3
- *Infrared Fiber Optics*, Paul Klocek, George H. Sigel, Jr., Vol. TT2
- *Spectrally Selective Surfaces for Heating and Cooling Applications*, C. G. Granqvist, Vol. TT1

Sensor and Data Fusion Concepts and Applications

Second Edition

Lawrence A. Klein

Tutorial Texts in Optical Engineering
Volume TT35

Donald C. O'Shea, Series Editor
Georgia Institute of Technology

SPIE Optical Engineering Press
A Publication of SPIE—The International Society for Optical Engineering
Bellingham, Washington USA

Library of Congress Cataloging-in-Publication Data

Klein, Lawrence A.
Sensor and data fusion concepts and applications / Lawrence A. Klein. — 2nd ed.
p. cm. – (Tutorial texts in optical engineering; v. TT35)
Includes bibliographical references and index.
ISBN 0-8194-3231-8
1. Signal processing—Digital techniques. 2. Multisensor data fusion. 3. Algorithms. I. Title. II. Series.
TK5102.9.K56 1999
621.382'2—dc21 99-14120
CIP

Published by

SPIE—The International Society for Optical Engineering
P.O. Box 10
Bellingham, Washington 98227-0010
Phone: 360/676-3290
Fax: 360/647-1445
Email: spie@spie.org
WWW: http://www.spie.org/

Printed in the United States of America.

Third Printing

To Amy, Gregory, and Jonathan

Introduction to the Series

The Tutorial Texts series was begun in response to requests for copies of SPIE short course notes by those who were not able to attend a course. By policy the notes are the property of the instructors and are not available for sale. Since short course notes are intended only to guide the discussion, supplement the presentation, and relieve the lecturer of generating complicated graphics on the spot, they cannot substitute for a text. As one who has evaluated many sets of course notes for possible use in this series, I have found that material unsupported by the lecture is not very useful. The notes provide more frustration than illumination.

What the Tutorial Texts series does is to fill in the gaps, establish the continuity, and clarify the arguments that can only be glimpsed in the notes. When topics are evaluated for this series, the paramount concern in determining whether to proceed with the project is whether it effectively addresses the basic concepts of the topic. Each manuscript is reviewed at the initial state when the material is in the form of notes and then later at the final draft. Always, the text is evaluated to ensure that it presents sufficient theory to build a basic understanding and then uses this understanding to give the reader a practical working knowledge of the topic. References are included as an essential part of each text for the reader requiring more in-depth study.

One advantage of the Tutorial Texts series is our ability to cover new fields as they are developing. In fields such as sensor fusion, morphological image processing, and digital compression techniques, the textbooks on these topics were limited or unavailable. Since 1989 the Tutorial Texts have provided an introduction to those seeking to understand these and other equally exciting technologies. We have expanded the series beyond topics covered by the short course program to encompass contributions from experts in their field who can write with authority and clarity at an introductory level. The emphasis is always on the tutorial nature of the text. It is my hope that over the next few years there will be as many additional titles with the quality and breadth of the first ten years.

Donald C. O'Shea
Georgia Institute of Technology

Contents

Figures

Tables

Preface to Second Edition

It became apparent through teaching a succession of short courses dealing with multisensor, multitarget data fusion techniques for tracking and identification of potential targets that explanations and examples of additional data fusion algorithms were needed. Three areas were of particular interest, namely, Bayesian inference, artificial neural networks, and fuzzy logic. This edition contains the material subsequently developed to satisfy the requests for information in these areas.

In addition, all of the original chapters in the first edition have been revised to varying degrees. The major additions are noted below. Chapter 2 on multiple sensor system applications, benefits, and atmospheric attenuation has been expanded to include information about the FASCODE and MODTRAN atmospheric models and EOSAEL, which analyzes a variety of physical processes that affect the performance of millimeter-wave and infrared sensors. Additional information has also been included about LOWTRAN, visibility metrics, and infrared propagation through snow. Where warranted, Chapter 3 has been expanded to provide explanations of data fusion architectures and algorithms. The chapter concerning the Dempster-Shafer algorithm has been revised to incorporate additional interpretations of the uncertainty interval. A second appendix has been added to discuss another type of confidence level architecture that is connected to voting fusion. Finally, the chapter about passive data association techniques for locating objects contains revisions concerning the application of the law of sines to target location and the branch-and-bound algorithm.

Martin Dana, with whom I taught the multisensor, multitarget data fusion course, reviewed several of the new sections. His insightful suggestions have improved upon the text. I want to thank Eric Pepper and Donald O'Shea of SPIE for giving me the opportunity to prepare the second edition of this volume. Sue Price, Rick Hermann, and the SPIE editorial staff provided many helpful comments that improved the quality of the text and assisted in its preparation.

Lawrence A. Klein

April 1999

Preface to First Edition

The material in this book was developed from a set of lecture notes on advanced sensor operation and data and sensor fusion. Because of space limitations commonly found in many applications of multisensor systems, the primary sensors examined were those operating at infrared and millimeter-wave frequencies. In addition to having relatively short wavelengths and hence occupying small volumes, these sensors have the ability to provide complementary information as they respond to different signature-generation phenomena. The result is a great degree of immunity to inclement weather, clutter, and signature masking such as from countermeasures.

Topics covered in the original lecture series included multiple sensor system applications to smart weapons; methods of allocating mission requirements among subsystems; signature-generation phenomena for active and passive sensors operating in different regions of the electromagnetic spectrum; radar systems operation including hardware issues, calculation of detection probability, analysis of target and clutter signatures, and descriptions of signal processing algorithms; infrared sensor design and performance trades for imaging and nonimaging sensors; laser radar operation and applications; impacts of jamming on sensor performance; and sensor and data fusion techniques that aid in target discrimination (detection, orientation, classification, and identification) and tracking.

Those topics that deal with applications of multiple sensor systems; target, background, and atmospheric signature-generation phenomena; and methods of combining multiple sensor data in target identity and tracking data fusion architectures were expanded for inclusion in this book. Most signature phenomena and data fusion techniques are explained with a minimum of mathematics or use simple mathematical operations to convey the underlying principles. Understanding of concepts is aided by the nonmathematical explanations that are provided in each chapter.

Weather forecasting, Earth resource surveys that use remote sensing, vehicular traffic management, target classification and tracking, and battlefield assessment are some of the applications that will benefit from the signature-generation phenomena and data fusion concepts developed in this book. Even with the deemphasis in some areas of defense spending, high interest continues in data fusion applications that assist in the identification of missile threats, other strategic targets, and the integration of potential responses. The signature-generation phenomena and fusion architectures and algorithms that are presented continue to be applicable to these areas and the growing number of nondefense applications. The final chapter that describes detecting and tracking objects using only passively acquired data should be of particular interest to those concerned with the missile location problem.

Several people have made valuable suggestions that were incorporated into this work. Henry Heidary, in addition to his major contributions to Chapter 9, reviewed other sections of the book. Sam Blackman reviewed the entire text and provided several references for new material that were subsequently incorporated. Pat Williams reviewed sections on tracking and provided data about tracking algorithm execution times. Mary Horan of SPIE performed the final editing in a superb manner. Finally, I want to thank my wife, Jane, and my children for supporting me in this endeavor by being understanding of my late-night stints at the computer terminal.

Lawrence A. Klein

April 1993

Sensor and Data Fusion Concepts and Applications

Second Edition

Chapter 1

Introduction

Weather forecasting, battlefield assessment, target classification and tracking, transportation management – these are but a few of the many civilian and defense applications that are performed using sensor and data fusion. Knowledge of target and background signature-generation phenomena, sensor design, signal processing algorithms, pertinent characteristics of the environment in which the sensors operate, available communications, and end use of the fusion products are generally used to optimize the size, cost, design, and performance of the sensor and associated data processing systems.

This book discusses the above topics, with primary emphasis on signature-generation phenomena to which electromagnetic sensors respond; atmospheric effects; and data fusion algorithms and architectures for target detection, classification, identification, and tracking. The types of signatures and data collected by a sensor are related to the following:

- Type of energy (e.g., electromagnetic, acoustic, ultrasonic, magnetic, seismic) received by the sensor;
- Active or passive sensor operation as controlled by center frequency, polarization, bandwidth, and incidence angle;
- Spatial resolution of the sensor versus target size;
- Target and sensor motion;
- Weather, clutter, and countermeasure effects.

Although the focus of this book is on phenomena that affect electromagnetic sensors, acoustic, ultrasonic, magnetic, and seismic sensors can also be a part of a sensor fusion architecture. The latter group of sensors has civilian uses in detecting vehicles on roadways and aircraft on runways as well as in geology. Military applications of these sensors include the detection and classification of targets above and below ground. The information that nonelectromagnetic sensors provide can certainly be part of a data and sensor fusion architecture.

Once the signature-generation processes or observables are known, it is possible to design a multiple sensor system that captures their unique attributes. Sensors that respond to signatures generated by different physical phenomena can subsequently be selected and their outputs combined to provide varying degrees of immunity to weather, clutter, and diverse countermeasures. Oftentimes, the data fusion process produces knowledge that is not otherwise obtainable or is more accurate than information gathered from single sensor systems. An example of the former is the identification of vegetation on Earth through fusion of hyperspectral data from spacebased sensors such as the Airborne Visible/Infrared Imaging Spectrometer (AVIRIS). The AVIRIS contains 224 detectors, each with a spectral bandwidth of approximately 10 nm that cover the 380- to 2500-nm band. Data fusion also improves the ability of missiles to track and defeat threats. In this case, accuracy is enhanced by handing off the guidance required for final missile impact from a lower resolution sensor, optimized for search, to a higher resolution sensor optimized to find a particular impact area on a target.

The definition of data fusion that appears in this book is derived from recommendations of the U.S. Department of Defense Joint Directors of Laboratories Data Fusion Subpanel, namely,

> *Data fusion is a multilevel, multifaceted process dealing with the automatic detection, association, correlation, estimation, and combination of data and information from multiple sources.*

Data fusion consists of a collection of subdisciplines, some of which are more mature than others. The more mature techniques, such as Bayesian inference and multisensor, multitarget tracking, draw on a theoretical apparatus that supports their application. The less mature ones are dominated by heuristic and ad hoc methods.

The fusion architecture selected to combine sensor data depends on the particular application, sensor resolution, and the available processing resources.

- Application: For automatic target recognition, sensors may be allowed more autonomy in processing their data than if target tracking is the goal. Largely autonomous sensor processing can also be used to fuse the outputs of existing sensors not previously connected as part of a fusion architecture. Many target tracking applications, however, produce more reliable estimates of tracks when unprocessed multiple sensor data are combined and then used to identify new tracks or to correlate with existing tracks.

- Sensor resolution: If the sensors can resolve multiple pixels (picture elements) on the target of interest, then the sensor data can be combined pixel-by-pixel to create a new fused information base that can be analyzed for the presence of objects of interest. In another method of analysis, features can be (1) extracted from each sensor, (2) combined to form a new, larger feature vector, and (3) consequently input, for example, to a statistical-based algorithm to determine the object's classification.

- Processing resources: Individual sensors can be used as the primary data processors when sufficient processing resources are localized in each sensor. In this case, preliminary detection and classification decisions made by the sensors are sent to a fusion processor for final resolution. If the sensors are dispersed over a relatively large area and high data rate and large bandwidth communications media capable of transmitting unprocessed data to a central processing facility are in place, then a more centralized data processing and fusion approach can be implemented.

The following chapter describes the signature-generation phenomena and benefits associated with multiple sensor systems. The remaining chapters discuss sensor and data fusion signal processing architectures and algorithms suitable for automatic target recognition, target track estimation, and situation and threat assessment. The Bayesian, Dempster-Shafer, voting logic, and fuzzy logic data fusion algorithms that are discussed in some detail have one characteristic in common. They all require expert knowledge or information from the designer to define the respective *a priori* probabilities and likelihood functions, probability mass, confidence levels, or membership functions and production rules used by the algorithms. Other algorithms, such as classical inference, knowledge-based, and pattern recognition, also require the designer to assume probability density functions, rules, or other parameters for their operation. Implementation of the data fusion algorithms is thus dependent on the expertise and knowledge of the designer, analysis of the operational situation, and the types of information provided by the sensor data.

Chapter 2 illustrates the benefits of multiple sensor systems in locating and tracking targets in inclement weather, high clutter, and countermeasure environments. The attributes of the atmosphere, background, and targets that produce the signatures detected by electromagnetic active and passive sensors are described, as are models used to calculate the absorption, scattering, and propagation of millimeter-wave and infrared energy through the atmosphere.

Chapter 3 explores sensor and data fusion architectures and examines the different classes of data fusion algorithms applicable to automatic target detection, classification, and track estimation. The methods used to categorize data fusion are described, including those based on (1) where the sensor data are processed and fused and (2) the degree of processing that precedes the fusion of the data. The applications and implications of the taxonomies are discussed.

Chapter 4 presents a discussion of Bayesian inference, a statistical-based data fusion algorithm. Bayesian inference, based on Bayes' rule, is a method for calculating the conditional or *a posteriori* probability of a hypothesis being true given supporting evidence. By contrast, classical inference gives the probability that an observation can be attributed to an object or event, given an assumed hypothesis. A two-sensor mine detector is described to illustrate how data are combined through Bayesian inference.

Chapter 5 discusses the Dempster-Shafer algorithm, in which sensors contribute detection or classification information to the extent of their knowledge. One of the important features of the Dempster-Shafer approach is the ability to assign a portion of a sensor's knowledge to uncertainty, that is, the class of all events that make up the decision space. Dempster-Shafer differs from Bayesian inference in this respect.

Chapter 6 examines artificial neural networks and the algorithms commonly used to train linear and nonlinear single and multilayer networks. The supervised training paradigms include minimization of the least mean square error between the known input and the learned output, the perceptron rule, and the backpropagation algorithm that allows the weights of hidden layer neurons to be optimized. Other nonlinear training algorithms and neural networks that use unsupervised learning are described as well.

In Chapter 7, a voting algorithm derived from Boolean algebra is discussed. Here each sensor processes the information it acquires using algorithms tailored to its resolution, scanning, and data processing capabilities. The outputs from each sensor are input to the fusion algorithm, where series and parallel combinations of the sensor outputs are formed and a decision is made about an object's classification.

Chapter 8 describes fuzzy logic and fuzzy neural networks. Fuzzy logic is useful when input variables do not have hard boundaries or the exact mathematical formulation of a problem is unknown. Fuzzy logic may also decrease the time needed to compute a solution when the problem is complex and multi-dimensional. In fuzzy set theory, an element's membership in a set is a matter of degree and an element may be a member of more than one set. Fuzzy logic permits control statements or production rules, also called fuzzy associative memory, to be written to accommodate the imprecise states of the variables. Several types of defuzzification operations are discussed to convert the output fuzzy values into a fixed and discrete output that is utilized by the control system. The balance of an inverted pendulum and track estimation with a Kalman filter are two examples used to illustrate the wide applicability of fuzzy logic. Adaptive fuzzy neural systems are also discussed. These rely on sample data and neural algorithms to define the fuzzy system at each time instant.

Chapter 9 examines three fusion architectures suitable for fusing passively acquired data to locate and track targets that are emitters of energy. This material was written, in part, by Henry Heidary of Hughes Aircraft Company, now Raytheon Systems Company. In theory, any form of emitted energy (microwave, infrared, visible, acoustic, ultrasonic, magnetic, etc.) can be located with the proper array of passive receivers. These three approaches permit the range to the emitters to be estimated using only the passively received data. Two of the architectures use centralized fusion to locate the emitters. One of these analyzes the unprocessed received signal waveforms, while the other associates azimuth and elevation angle measurements to estimate the location of the emitters. The third architecture uses a distributed processing concept to associate the angle tracks of the emitters that are calculated by the individual sensors. Factors that influence the signal processing and communications requirements imposed by each of the methods are discussed.

Chapter 2

Multiple Sensor System Applications, Benefits, and Atmospheric Attenuation

Millimeter-wave (MMW) and infrared (IR) sensors are frequently used in autonomous multiple sensor systems because their operating frequencies allow relatively compact designs to be realized. Objects that may be difficult to discriminate among when only a single sensor is used are often differentiated with a multiple sensor system that exploits several signature-generation phenomena to gather data about the objects and scene of interest. Relatively high probabilities of object detection and classification, at acceptable levels of false classification, are achieved in inclement weather, high-clutter, and counter-measure environments by using operating frequencies that cover a wide portion of the electromagnetic spectrum. Signatures generated by multiple phenomena also expand the amount of information that can be gathered about the location of vulnerable areas on targets. This is important in smart munition applications where autonomous sensors guide weapons to their targets without operator intervention. Multiple sensor systems are used in civilian applications as well, such as space-based sensors for weather forecasting and Earth resource surveys. Here narrow-band wavelength spectra and multiple receiver configurations, such as active radar transmitters, passive radar receivers, and infrared and visible sensors, provide data about temperature, humidity, rain rates, wind speed, storm tracks, snow and cloud cover, and crop type and maturity. Acoustic, ultrasound, magnetic, and seismic signature-generation phenomena are also used in military and civilian applications, but these are not addressed in this book. However, their data can be fused with those of other sensors using the algorithms and architectures described in later chapters.

A *sensor* consists of front-end hardware, called a transducer, and a data processor. The transducer converts the energy entering the aperture into lower frequencies from which target and background discrimination information is extracted in the data processor. A *seeker* consists of a sensor to which scanning capability is added to increase the field of regard. Seekers may take the form of sensors placed on single- or multiple-axis gimbals, IR detector arrays illuminated by scanning mirrors that reflect signals from a large field of regard, or frequency sensitive antenna arrays whose pointing direction changes as the transmitted frequency is swept over some interval.

2.1 Data fusion applications to multiple sensor systems

One application of sensor and data fusion is in smart munitions that use multiple sensor data to precisely guide warheads and missiles to the desired targets by providing real-time tracking and object classification information, while simultaneously minimizing risk or injury to the personnel launching the weapon. Other applications include tracking of aircraft and missiles with multiple sensors located on spatially separated platforms (ground, air, sea, or space based, or in any combination) or with multiple sensors that are collocated. Spatially separated sensor locations reduce the number of time intervals where targets are blocked from the view of any of the sensors, making tracking data available for larger portions of the target's flight time. The process of combining tracks produced by the sensors involves fusion of the data. When collocated multiple sensors are used, a sensor having a large field of view may be employed, for example, to search a large area. A portion of this area may then be handed off and searched with higher resolution sensors to obtain more accurate tracking or object identification data in the restricted region of interest. The process of conveying the location of the restricted search area to the higher resolution sensor makes use of fusion functionality.

Multiple sensors, responding to different signature-detection phenomena, may also be used to increase the probability that a target signature will be found during a search operation. Objects that may not be recognizable to one sensor under a given set of weather, clutter, or countermeasure conditions may be apparent to the others. Another application of sensors that respond to independent signature-generation phenomena is exemplified by a radar supplying range data to a higher resolution infrared sensor that lacks this information. By properly selecting signal processing algorithms that combine the range data with data from the infrared sensor, new information is obtained about the absolute size of the objects in the field of view of the sensors. The process of combining the multisensor data again involves data fusion.

Functions that sensors perform in precision guided weapons applications are summarized in Table 2.1. They are implemented with hardware, software, or combinations of both. Sensor fusion is implicit when multiple sensor data are used to support a function. These sensor functions, with the exception of warhead firing or guidance, carry over into nonmilitary applications. For example, in some intelligent transportation system applications, it is necessary to detect, classify, and track vehicles in inclement weather (such as rain and fog) where the signature contrast between vehicle and background may be reduced or the transmitted energy attenuated.

In addition to the applications discussed above, multiple sensors are used for weather forecasting and Earth resources measurements. Weather satellites rely on combinations of microwave, millimeter-wave, infrared, and visible sensors to gather data about temperature and water vapor profiles through the atmosphere, rain rates, cloud coverage, storm tracks, sea state, snow pack, and wind velocities, to name a few. These applications require the reception of data at as many frequencies and polarizations, or any combination thereof, as there are meteorological parameters to calculate. The parameters are then ascertained by inverting the equations containing the measured data and the parameters of interest.[1,2,3,4,5,6,7] Satellites such as LANDSAT use visible and IR wavelengths to provide information about crop identity and maturity, disease, and acreage planted. Synthetic aperture radar (SAR) is used in still other spacecraft to provide imagery of the Earth, even through cloud cover.[8] SAR provides yet another source of space-based information that can be fused with data from other sensors.

Table 2.1: Common sensor functions and their implementations in precision guided weapons applications

Function	Implementation
• Target detection	• Multiple threshold levels (may be bipolar)
	• Data processing
• False alarm and false target rejection	• Data processing
• Target prioritization	• High-resolution sensors
	• Object classification algorithms
• Countermeasure resistance	• Control of transducer apertures – Antenna beamwidth and sidelobes – IR pixel size (instantaneous field of view)
	• Receive multiple signatures generated by independent phenomena
	• Data processing
• Target tracking	• Seeker hardware
	• Algorithms that fuse tracks and data from multiple sensors and multiple targets
• Warhead firing or guidance command to hit desired aimpoint	• Fine spatial resolution sensors
	• Data processing

2.2 Selection of sensors

Data acquired from multiple sensor systems are more likely to be independent when the operating frequencies of the sensors are selected from as wide an expanse across the electromagnetic spectrum as is possible and, furthermore, when they are used in both active (transmit and receive) and passive (receive only) modes of operation as indicated in Figure 2.1. Examples of active sensors are MMW and laser radars. Examples of passive sensors include MMW radiometers and IR radiometers, FLIR (forward looking infrared) sensors, and IRST (infrared search and track) sensors. In selecting the operating frequencies or wavelengths, tradeoffs are frequently made among component size; resolution; effects of weather, atmosphere, clutter, and countermeasures; and cost. For example, a microwave radar operating at a relatively low frequency is comparatively unaffected by the atmosphere (especially for shorter-range applications), but can be relatively large in size and not provide sufficient resolution. A higher-frequency radar, while smaller in size and of better resolution for the same size aperture, may be higher in cost and more susceptible to atmospheric and weather effects.

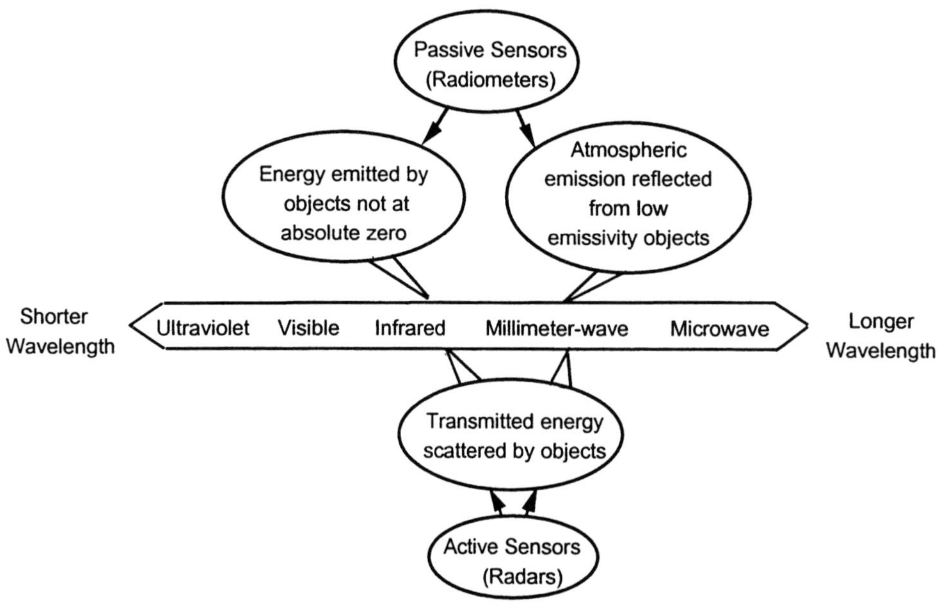

Figure 2.1: Signature-generation phenomena in the millimeter-wave and infrared electromagnetic spectrum

Sensors designed for weather forecasting operate at frequencies where energy is either known to be absorbed by specific molecules (such as oxygen to provide atmospheric temperature profiles or water to provide water vapor profiles) or at frequencies at which the atmosphere is transparent in order to provide measurements at the Earth's surface or at lower altitudes. Other applications, such as secure communications systems, may operate at a strong atmospheric absorption frequency, such as the 60-GHz oxygen complex, to prevent transmission over long distances and to make interception of the information difficult.

Radar sensors operate within frequency bands that are identified by the letter designations shown in Table 2.2. Frequencies in K band and below are usually referred to as microwave and those at Ka band and above as millimeter wave.

Table 2.2: Radar spectrum letter designations

Letter	Frequency (GHz)	Free Space Wavelength (mm)
L	1 to 2	300 to 150
S	2 to 4	150 to 75.0
C	4 to 8	75.0 to 37.5
X	8 to 12	37.5 to 25.0
Ku	12 to 18	25.0 to 16.6
K	18 to 26.5	16.6 to 11.3
Ka	26.5 to 40	11.3 to 7.5
Q	33 to 50	9.1 to 6.0
U	40 to 60	7.5 to 5.0
V	50 to 75	6.0 to 4.0
E	60 to 90	5.0 to 3.3
W	75 to 110	4.0 to 2.7
F	90 to 140	3.3 to 2.1
D	110 to 170	2.7 to 1.8
G	140 to 220	2.1 to 1.4

IR sensors operate over spectral regions in the near-, mid-, and long-wavelength IR spectral bands that correspond roughly to 0.77 to 1.5 μm, 1.5 to 6 μm, and 6 to 40 μm, respectively. These bands are usually restricted even further with spectral filters to maximize the response to particular object or molecular signatures and eliminate false returns from the surrounding atmosphere and background.

Active sensors such as MMW radars operate in monostatic and bistatic configurations. In the monostatic mode, the transmitter and receiver are collocated and the receiver processes energy that is backscattered from objects in the field of view. In the bistatic mode, the transmitter and receiver are spatially separated. Here the receiver processes energy that is scattered toward it by objects, as shown in Figure 2.2. When the bistatic angle β is equal to zero, the configuration reverts to the monostatic case. Bistatic radars do not enjoy as many applications as monostatic radars. They do find use, however, in applications requiring detection and tracking of stealth targets, air-to-ground attack scenarios, satellite tracking, semiactive tracking of missiles, and passive situation assessment.[9]

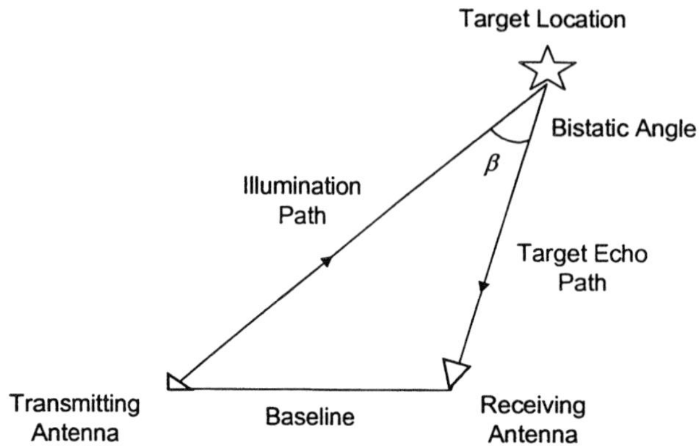

Figure 2.2: Bistatic radar geometry

In the monostatic and bistatic MMW radar configurations, the received signal contains information about scatterer size and location as illustrated in Figure 2.3. IR laser radars provide similar information but at higher resolution, due to their shorter wavelength. However, IR laser radars are subject to greater atmospheric attenuation and an inability to search large areas in a short time. In addition to scatterer size, shape, and location, the energy received by laser radar is also responsive to the differences in reflectance between the objects and their backgrounds. This added discriminant can assist in differentiating targets from backgrounds and other objects.[10]

The inverse relation of sensor resolution to wavelength is depicted in Figure 2.4. In this illustration, the apertures and effective range of the sensors are kept constant at 8 inches (20 cm) and 5 km, respectively, as the operating frequency varies from microwave through visible.

IR passive sensors, such as radiometers, respond to the apparent temperature difference between target and background. The apparent temperature depends on the absolute temperature of the object in the field of view of the radiometer and on the emissivity of the object in the IR spectral band of interest. Temperature sources in the sensor itself that emit energy into the aperture of the sensor also affect the apparent temperature. FLIR and IRST sensors are other types of passive IR devices. FLIRs are primarily used to provide high-resolution imagery of a scene, while IRSTs are primarily used to locate a "hot" area on an object and thus track it. Design parameters that optimize the performance of FLIRs, such as a small instantaneous field of view, may hinder the performance of IRSTs that require a small noise-equivalent temperature difference.[11,12] Accordingly, one sensor design may not be optimal for all applications. Millimeter-wave radiometers, not shown in the figure, behave in a similar manner to the IR radiometer. They respond to the absolute temperature of the object and its emissivity at the MMW operating frequency of the receiver. Since metal objects

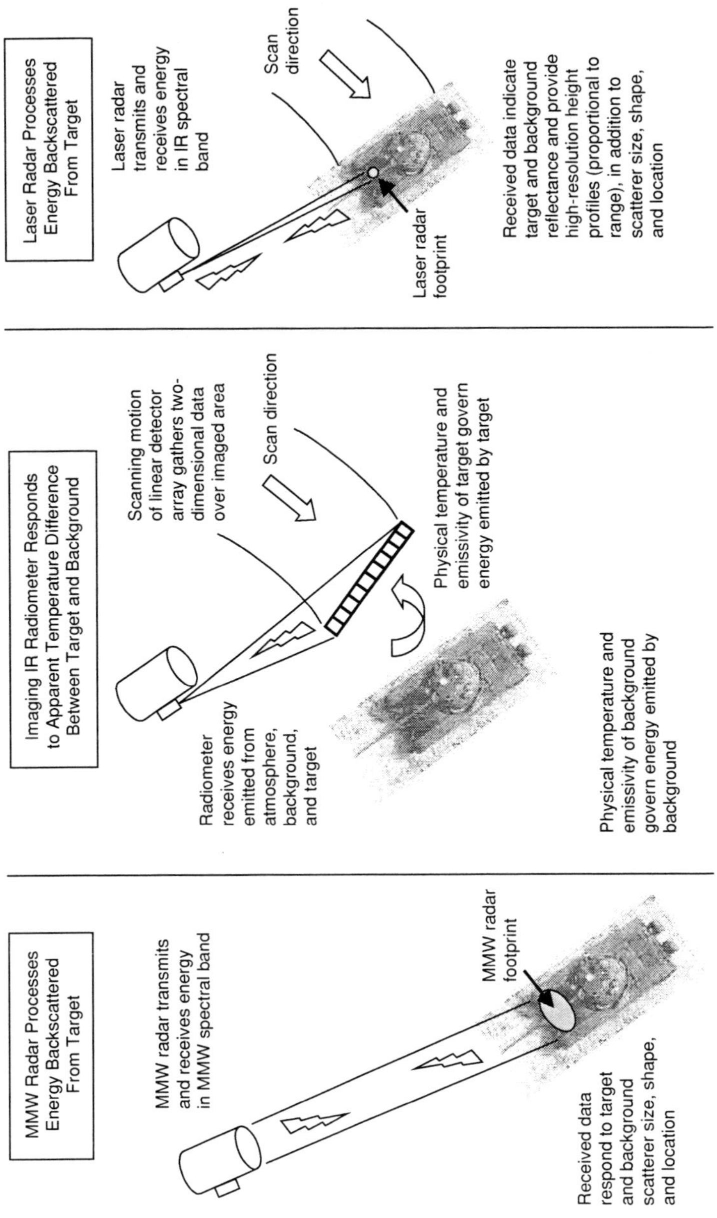

Figure 2.3: Active and passive sensors operating in different regions of the electromagnetic spectrum produce target signatures generated by independent phenomena

have low emissivity and hence high reflectivity at MMW frequencies, their passive signatures are mainly due to (1) reflection of the downwelling component of atmospheric energy from the metal and (2) emission from the upwelling component of energy produced by the background and atmosphere as described by radiative transfer theory in Appendix A.

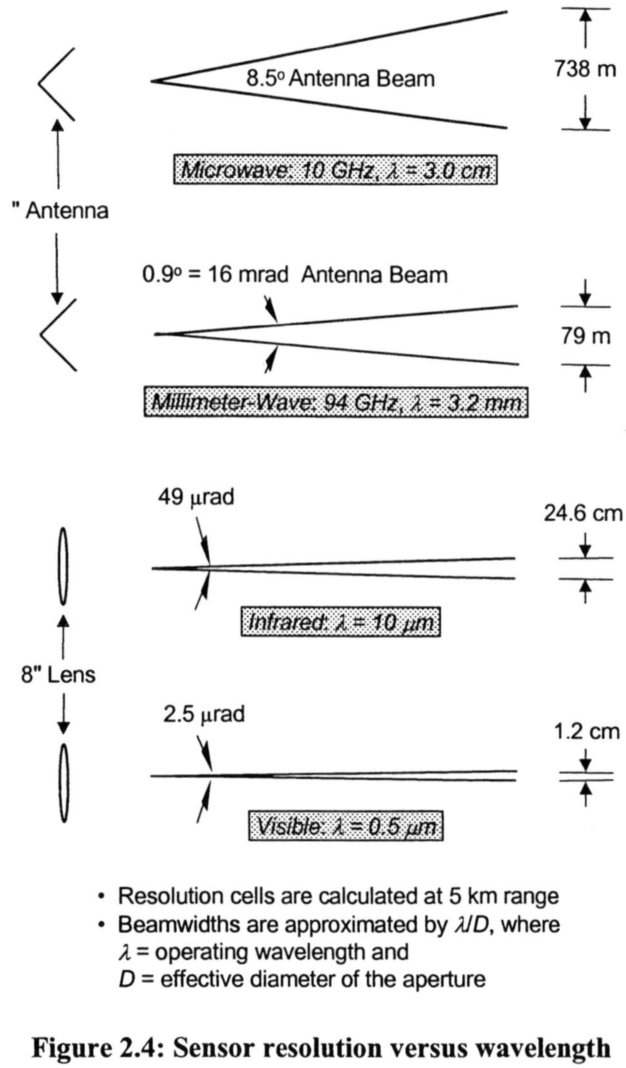

Figure 2.4: Sensor resolution versus wavelength

Cost and sensor performance goals in military applications are influenced by the value of the target the sensor helps defeat. Sensors designed to defeat low-value targets, such as tanks, trucks, and counterfire batteries, are generally of low cost (several thousand to tens of thousands of dollars), while sensors designed for high-value targets such as aircraft, ships, and bridges can cost hundreds of thousands of dollars. One of the goals of multiple sensor systems is to reduce the cost of smart munitions and tracking systems, whether for the low- or high-value target. This can be achieved by using combinations of lower cost sensors, each of which responds to different signature-generation phenomena, to obtain target classification and tracking data previously available only with expensive, single phenomena-oriented sensors. Modern missiles may also incorporate Global Positioning System (GPS) receivers to update their trajectory by fusing the GPS data with data from the onboard sensors.

An example of a multiple sensor system that can support automatic target recognition (ATR) is shown in Figure 2.5. For illustration, MMW radar, MMW radiometer, and passive and active IR sensors are shown. In this sensor-level fusion configuration, each sensor processes its data with algorithms that are tailored and optimized to the received frequency band, active or passive nature of the sensor, spatial resolution and scanning characteristics, target and background signatures, polarization information, etc. Results of the individual sensor processing are forwarded to a fusion processor where they are combined to produce a validated target or no-target decision.

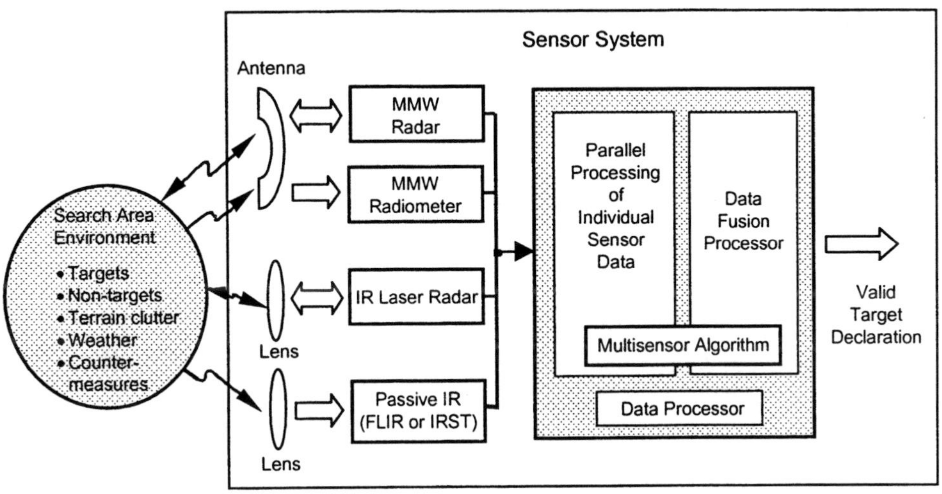

Figure 2.5: Sensor fusion concept for ATR using multiple sensor data

If target track estimation is the desired output of the multiple sensor system, then another method of combining the sensor data proves to be more optimal in producing accurate tracks in many applications. In this configuration, called central-level fusion, minimally processed sensor data are associated in the fusion processor. Correlated data are combined to form tracks and estimate future positions of the targets as explained in Chapter 3.

2.3 Benefits of multiple sensor systems

A quantitative argument can be made for the use of multiple sensor systems as illustrated in Figure 2.6. The lower curve gives the detection probability for a single radar sensor as a function of signal-to-noise ratio when the false alarm probability is equal to 10^{-6}. The detection probability of 0.7 is adequate when the signal-to-noise ratio is nominal at 16 dB. But when the target signature is reduced and the signal-to-noise ratio decreases to 10 dB, the detection probability falls to 0.27, generally not acceptable for radar sensor performance.

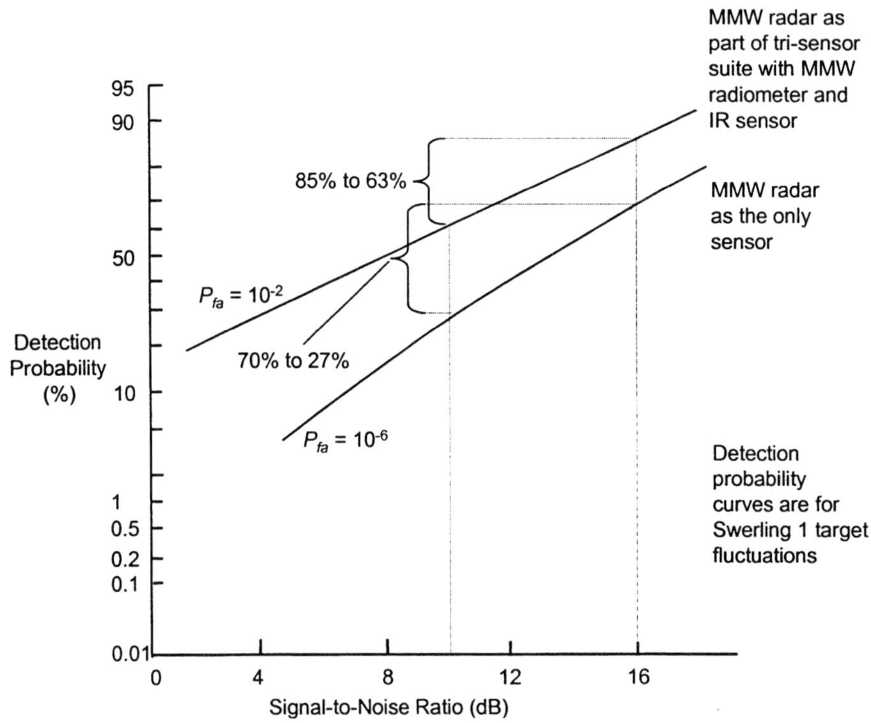

Figure 2.6: Multiple sensor versus single sensor performance with suppressed target signatures

If, however, the radar is one of three sensors that detect the target, where each sensor responds to unique signature-generation phenomena and does not generally false alarm on the same events as the others, then the false alarm rejection can be distributed among the three sensors. The system false alarm probability of 10^{-6} is recovered later in the fusion process when the data are combined, for example, with an algorithm such as voting fusion that incorporates sensors operating in series and parallel combinations. When the false alarm rejection can be divided equally among the sensors, the radar performance is given by the upper curve marked with the 10^{-2} false alarm probability. Now the nominal target signature yields a detection probability of 0.85, but even more importantly, the reduced-signature target (with signal-to-noise ratio 10 dB) yields a detection probability of 0.63, which is two and a third times greater than before. Thus, multiple sensors allow the false alarm rejection to be spread over the signature acquisition and signal processing capabilities of all the sensors and the data combining capabilities of the fusion algorithm. This architecture potentially lets each sensor operate at a higher false alarm probability and increases the detection probability of the sensors especially when target signatures are suppressed.

An example of the object discrimination capabilities provided by combining active and passive MMW sensor data is shown in Figure 2.7. Examination of the truck top and shingle roof signatures (on the left of the figure enclosed by dashed lines) shows that it is difficult to tell whether the object is a truck or a roof with only radar data, as both have about the same radar cross section and, hence, relative radar backscatter returns. If a radiometer is added to the sensor mix, the difference in the two objects' signatures stands out as shown on the vertical target/background temperature contrast scale. But, if only a radiometer is available, it is difficult to discern an asphalt road from a truck as shown in the dashed region on the right of the figure. However, the radar now adds the discriminating signatures, making object differentiation possible.[13]

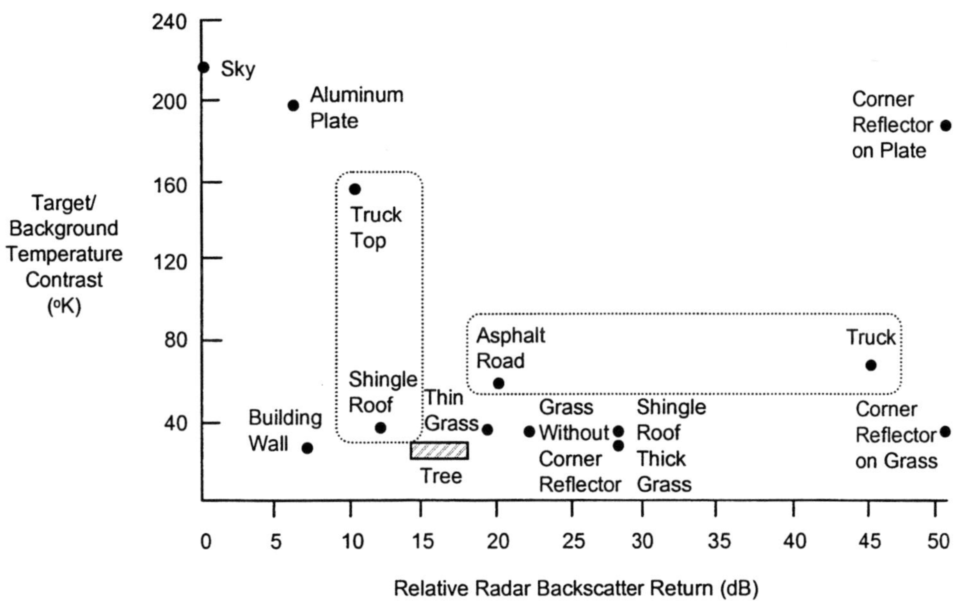

Figure 2.7: Target discrimination with MMW radar and radiometer data

Multiple sensors also act in a synergistic manner when operating in high-clutter environments and in inclement weather. A sensor, such as MMW radar, that may be hampered by the high clutter of dry snow is aided in detecting targets by a passive sensor that is not similarly affected. But, an IR sensor that may be impaired by dust or clouds is augmented by the MMW radar in detecting targets under these conditions.

Another example of sensor synergy occurs through the information multiple sensors provide about the location of a potential target's vulnerable area. A passive MMW radiometer supplies data to compute the centroid of the object that can be used as a potential aimpoint location. A high-resolution FLIR can provide data to locate the boundary of an object and a region of warmer temperatures within that area. With suitable knowledge about the targets, the warmer region can be inferred to belong to the area over the engine, which is an ideal aimpoint. This imagery, as well as the passive MMW centroid data, allows the aimpoint to be located within the boundary of the object and avoids the pitfalls of simple hot-spot detection, which can declare a "false aimpoint" (e.g., from tracking hot exhaust gases) located outside the physical area of the target.

Benefits from multiple sensor systems also accrue from their ability to defeat countermeasures deployed to make a sensor ineffective either by jamming or by mimicking target signatures that deflect a sensor-guided missile away from the true target track. Multiple sensors either completely or partially defeat these countermeasures by exploiting target signature phenomena that are not countered or by driving up the cost of the countermeasure by requiring it to be more complex to replicate target signatures over a wide spectral band in the active and passive signature domains.

2.4 Influence of wavelength on atmospheric attenuation

Atmospheric attenuation is produced by two phenomena — absorption and scattering. Absorption is dependent on the frequency of operation and the gases and pollutants that are present. Scattering is dependent on the size, shape, and dielectric constant of the scattering objects and the wavelength of the incident energy. Atmospheric constituents such as oxygen, water vapor, and carbon dioxide play a dominant role in determining MMW and IR attenuation. The internal energy states of these molecules define frequencies at which the molecules absorb energy, thus creating frequency bands of high attenuation. These regions of the electromagnetic spectrum may be used to broadcast short-range communications that are intended to be difficult to intercept and to gather information used for weather forecasting and cloud top location. Relatively low absorption exists at still other portions of the electromagnetic spectrum called windows. Sensors that operate at these frequencies can propagate energy over greater distances for long-range target detection and for Earth resource monitoring. Weather-related obscurants such as rain, fog, and snow add to the absorption and scattering experienced under clear weather conditions and further limit sensor performance. Models that adequately predict atmospheric absorption and scattering in the MMW and IR spectra may be used when measured data are not readily available at specific frequencies or atmospheric conditions. In the microwave and millimeter-wave portions of the electromagnetic spectrum, atmospheric attenuation generally increases as the operating frequency increases.

In the infrared portion, attenuation is a strong function of the gases and pollutants that are present.

The higher resolution IR and visible sensors suffer greater performance degradation from the atmosphere, as seen in Figure 2.8. The curve with the peaks and valleys in attenuation corresponds to 1 atm of pressure at a temperature of 20°C and water density of 7.5 g/m^3. The window frequencies in the MMW spectrum, denoted by absorption minima, occur at approximately 35, 94, 140, 225, and 350 GHz. These windows are the frequencies and wavelengths typically used in sensors designed to detect potential targets. Peak absorption occurs in the microwave and millimeter-wave spectra at approximately 22, 60, 118, 183, and 320 GHz. Absorption at 60 and 118 GHz is due to oxygen, while absorption at the other frequencies is due to water vapor.

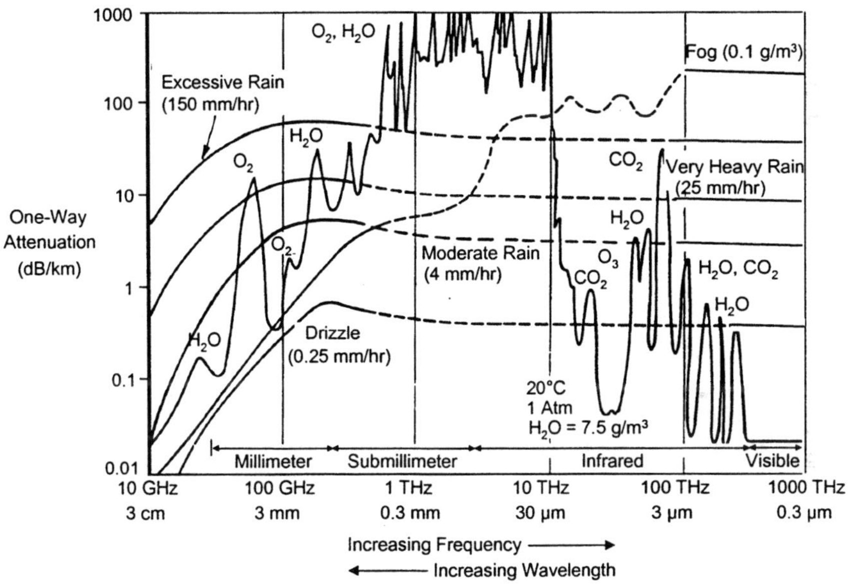

Figure 2.8: Atmospheric attenuation spectrum from 0.3 μm to 3 cm

The infrared absorption spectra are due to molecular rotations and vibrations that occur in atmospheric molecules. The near-IR wavelength band extending from 0.77 to 1.5 μm is constrained at the upper end by water vapor absorption. The mid-IR wavelength band from 3 to 5 μm is bounded on the lower and upper ends by water vapor absorption. An absorption peak in the middle of the band is due to carbon dioxide. The far-IR band or thermal IR extends from approximately 8 to 12 μm and beyond. The lower wavelength is restricted by water vapor and the upper by a combination of water vapor and carbon dioxide.

Rain rates of 0.25, 4, 25, and 150 mm/hr affect attenuation as shown in the curves in Figure 2.8. At frequencies below approximately 100 GHz, drizzle (0.25 mm/hr) produces less attenuation on MMW than on IR. In moderate and heavier rain, MMW frequencies of 97 GHz and above are generally subject to similar attenuation as the near IR as the rain rate curves of 4, 25, and 150 mm/hr show. Figure 2.8 also shows that a fog with 0.1 g/m^3 liquid water content is a greater attenuator of IR energy than MMW energy. Additional data describing the effects of water, in the form of rain and fog, on the propagation of MMW and IR energy are discussed in subsequent sections. Other atmospheric constituents, such as carbon dioxide, carbon monoxide, nitrous oxide, oxygen, methane, and ozone are treated by the computer models described in Section 2.14.

Propagation of visible, IR, and MMW energy through snow was studied during the Snow-One and Snow-One-A experiments conducted by the U.S. Army Cold Regions Research and Engineering Laboratories (CRREL) in 1981 and 1982. Transmittance and attenuation data are found in their report and other sources.[14,15] Table 2.3 gives the model for the extinction coefficient for mid- and far-infrared propagation through snow that was developed by Seagraves and Ebersole using these data.[16] They found that the extinction coefficient could be expressed as a function of only the visible extinction coefficient when the relative humidity was less than or equal to 94 percent. When the relative humidity was larger, making the occurrence of fog more likely, the infrared extinction coefficient was found a function of temperature and humidity as well.

Table 2.3: Extinction coefficient model for snow
(M.A. Seagraves, and J.F. Ebersole, "Visible and infrared transmission through snow," *Optical Engineering*, Vol. 22, No. 1, 90-93 [Jan.-Feb. 1983])

Applicable Wavelength	Applicable Humidity	Extinction Coefficient Model
3.0 μm	≤ 94 percent	$\gamma_{3.0} = 1.21\gamma_{0.55}$ Np/km
3.0 μm	> 94 percent	$\gamma_{3.0} = \gamma_{0.55}(-0.107T - 0.101H - 0.042V_i + 10.74)$ Np/km
10.4 μm	≤ 94 percent	$\gamma_{10.4} = 1.18\gamma_{0.55}$ Np/km
10.4 μm	> 94 percent	$\gamma_{10.4} = \gamma_{0.55}(-0.182T - 0.223H - 0.426V_i + 25.35)$ Np/km

The parameters that appear in the model are defined as

$\gamma_{0.55}$ = extinction coefficient at visible wavelengths (0.55 μm),

$$\frac{\gamma_{0.55}}{V_c} = 0.0233 - 0.0031V_c - 0.0101T + 0.0019H \text{ Np/km}, \tag{2-1}$$

V_c = volume concentration of snow in 10^{-8} m^3/m^3

$$= R/v\,, \tag{2-2}$$

R = equivalent liquid water precipitation rate,

v = particle settling velocity,

T = surface temperature in °C,

H = surface relative humidity in percent, and

V_i = visibility in km

$$= \frac{3.0}{\gamma_{0.55}}. \tag{2-3}$$

Since the model was derived from data with visibility, temperature, and humidity values in the ranges 1.2 km $\leq V_i \leq$ 7.5 km, -11.9°C $\leq T \leq$ 2.0°C, and 68% $\leq H \leq$ 100%, respectively, it should be applied with caution elsewhere. The model produces the largest errors in transmittance as compared to measured data when the relative humidity is between 90 and 95 percent, probably because the presence of fog is most in doubt in this region.

2.5 Fog characterization

Fogs found over land are of two types, advective fog (formed by cool air passing over a colder surface) typical of coastal regions, and radiative fog (formed by radiative cooling of the Earth's surface below its dew point level) found in inland regions. Advective fogs contain a greater number of large water drops and generally higher liquid water content than radiative fogs.[17] When the size of a particle in fog, cloud, rain, dust, etc. is comparable to the wavelength of the incident energy, the phase of the wave is not uniform over the particle. These phase differences give rise to the observed scattering of energy. Therefore, when the ratio of particle size to wavelength approaches unity, the energy attenuation increases. Thus, attenuation of shorter wavelengths (higher frequencies) can be greater in advective fogs because of the greater number of large particles and because of the larger liquid water content of the fog.

Optical visibility is commonly used to characterize fog when MMW attenuation is measured. However, optical visibility is hindered by the Mie scattering[18] of light from droplets in the fog, whereas energy at MMW wavelengths is not.[1] Therefore, the propagation of millimeter-waves through fog may be significantly greater than it appears to the human eye. Although water density appears to be a more precise measure of fog characterization, the transient nature of a fog makes it difficult to obtain this measure. Hence, the optical visibility characterization persists in comparisons of energy propagation through fog for MMW and IR systems. Visibility metrics are discussed further in Section 2.10.

2.6 Effects of operating frequency on MMW sensor performance

Table 2.4 summarizes the relationship of operating frequency on MMW sensor antenna resolution, atmospheric attenuation, and hardware design parameters. At a fixed aperture, a higher operating frequency reduces the antenna beamwidth and increases resolution. The increased resolution, while increasing pointing accuracy and reducing clutter cell size, may adversely affect the ability to search large areas within an acceptable time. This is due to the inverse relationship between sensor resolution and field of view (higher resolution, smaller field of view), or equivalently, the direct relationship between resolution and scan rate (higher resolution implies higher scan rate to search a given area in the same allotted time). The relation of frequency to atmospheric attenuation has already been introduced through Figure 2.8. Measurement data and models for estimating absorption and scattering of MMW energy by rain and fog are described in the following sections.

Table 2.4: Influence of MMW frequency on sensor design parameters

Parameter	Effect of Higher Frequency
Aperture	Higher gain
Pointing accuracy	Smaller error (standard deviation)
Clutter cell size	Smaller
Attenuation in air	Generally higher
Attenuation in rain and fog	Generally higher
Power available	Generally less
Component size	Smaller
Receiver noise figure	Generally higher
Integrated components in production	Less likely

Average power outputs from GaAs IMPATT (Impact Avalanche and Transit Time) diodes operating at 10 GHz and Si IMPATT diodes operating at 100 GHz

[1]. Mie scattering theory gives the general solution for the scattering of electromagnetic waves by a dielectric sphere of arbitrary radius. Rayleigh scattering, a limiting case of Mie scattering, applies when the wavelength is much larger than the scatterer's diameter.

have increased approximately 3 dB/decade.[19] Receiver noise figures at 35 and 95 GHz are roughly equal, although they are dependent on the technology used to manufacture the mixer diodes. Noise figures are larger at still higher frequencies. Since the higher frequency technologies are newer and applications fewer, there are currently less of these components available in integrated circuit designs.

2.7 Absorption of MMW energy in rain and fog

Rain affects the propagation of millimeter waves through absorption and backscatter. Figure 2.9 gives the one-way absorption coefficients (in dB/km) for MMW propagation through rain and fog.[20,21,22,23] For two-way radar applications, the absorption coefficient is doubled and then multiplied by the range between transmitter and target to get the absorption in decibels by which the energy reaching the sensor is reduced. Figure 2.9a shows measured values of the absorption coefficient for 15.5, 35, 70, 94, 140, and 225 GHz as a function of rain rate. Measured absorption data in fog are difficult to gather because of the nonsteady-state character of a fog.

The measured absorption coefficients in rain are predicted from the theoretical model data shown in Figure 2.9b by the solid curves corresponding to rain rates of 0.25, 1, 4, and 16 mm/hr. The modeled value of absorption is calculated using the Laws and Parsons drop-size distribution corresponding to the rain rate.[24] This distribution contains the number of droplets with diameters of specific size (0.05 cm to 0.7 cm in increments of 0.05 cm) as a percent of the total rain volume for rain rates of 0.25 to 150 mm/hr. Crane[20,25] found that differences between calculated values of absorption obtained from the Laws and Parsons drop-size distribution and from a large number of observed drop-size distributions were not statistically significant for frequencies up through 50 GHz. At higher frequencies, the drop-size distribution measurement errors in the small drop-size range affected the accuracy of the absorption versus the rain-rate relationship. Therefore, effects produced by different droplet-size models could not be differentiated from effects due to absorption at these frequencies. The agreement of the modeled data with measured values allows the prediction of atmospheric absorption in rain over large regions of the millimeter-wave spectrum and rain-rate variation when measured values are lacking. The data in Figure 2.9b may by interpolated to obtain absorption for other values of rain rate.[22]

Because droplet diameters in fog are small compared with millimeter wavelengths, scattering loss is negligible when compared to absorption of millimeter-wave energy by a fog. The one-way absorption coefficient in fog has been modeled as a function of the volume of condensed water in the fog and the operating wavelength of the sensor.[22] The model gives the absorption K_α as

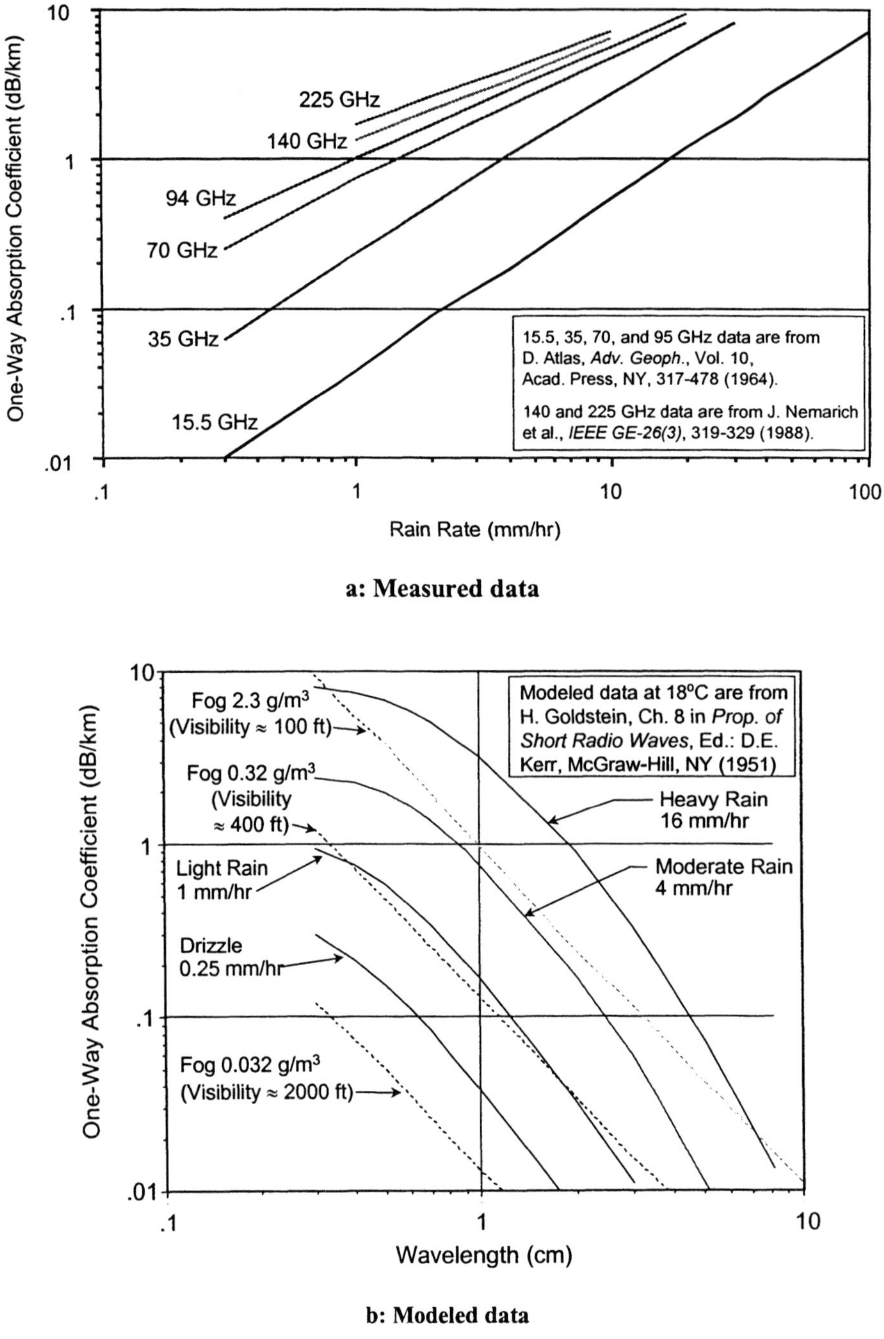

a: Measured data

b: Modeled data

Figure 2.9: Absorption coefficient in rain and fog as a function of operating frequency and rain rate or water concentration

$$K_\alpha = \frac{0.438 M_W}{\lambda^2} \text{ dB/km,} \tag{2-4}$$

where

K_α = one-way absorption coefficient,

M_W = mass of condensed water per unit volume of air in g/m^3, and

λ = sensor wavelength of operation in cm.

Equation (2-4) is accurate within 5 percent when 0.5 cm $\leq \lambda \leq$ 10 cm and when the droplets are extremely small with diameters of the order of 0.001 to 0.005 cm. A value of M_W = 1 g/m^3 represents about the maximum water content of most fogs, with the possible exception of heavy sea fogs. In most fogs, M_W is much less than 1. The FASCODE-1 weather model developed by the U.S. Air Force Geophysics Laboratory simulates[26] two heavy fogs with liquid water contents of 0.37 and 0.19 g/m^3 and two moderate fogs with liquid water contents of 0.06 and 0.02 g/m^3. (FASCODE is described further in Section 2.14.) For both types of simulated fog, the condensed water mass is less than 1. The modeled absorption data for fog, shown in Figure 2.9b by the dashed lines, are plotted from Eq. (2-4).

Ryde and Ryde, as reported by Goldstein, have given an empirical relation between an average $\overline{M_W}$ and optical visibility in fog, namely,[22]

$$\overline{M_W} = 1660 V_i^{-1.43} \tag{2-5}$$

where V_i is the optical visibility in feet and $\overline{M_W}$ is such that in 95 percent of the cases, M_W lies between 0.5 M_W and 2 M_W. Such a relation may be useful when more precise values of M_W are not available.

Calculations made by Richard et al.[27] show that there can be a difference of 8 dB/km in absorption at 140 GHz between advective and radiation fogs at 0.1-km visibility. Earlier measurements by Richer at the Ballistic Research Laboratories found a maximum one-way absorption of 23 dB/km at 140 GHz during a 30-s time period, that returned to a lower value of 15 dB/km during the following 30-s interval.[28] The change in absorption was not accompanied by an appreciable change in visibility. The measured 8-dB variation was attributed to an increase in fog density beyond the limits of human visibility or to the condensation of fog into rain along the propagation path.[29]

2.8 Backscatter of MMW energy from rain

Backscatter is a volumetric effect. Hence, the rain backscatter coefficient η (in m^2/m^3) is multiplied by the volumetric resolution cell V of the radar in cubic meters to obtain the equivalent radar cross section (RCS) of the rain in square meters. The rain RCS therefore acts as a "pseudotarget" and scatters energy toward the radar receiver that competes with the energy scattered from the real target.

The resolution cell V of the radar is given by

$$V = \pi/4\,(R\theta_{az})\,(R\theta_{el})\,(c\tau/2)\ \mathrm{m}^3, \tag{2-6}$$

where

R = range from the radar to the rain resolution cell in meters,

θ_{az}, θ_{el} = antenna 3-dB azimuth and elevation beamwidths, respectively, in radians,

τ = width of the transmitted pulse in seconds, and

c = speed of light in meters/second.

Thus, the RCS of the rain cell is given by

$$\mathrm{RCS} = \eta\, V\ \mathrm{m}^2. \tag{2-7}$$

If the range extent of the resolution cell is limited by a range gate of length L in meters, then $c\tau/2$ in Eq. (2-6) is replaced by L.

Rain backscatter coefficient data are shown in Figure 2.10 for linear polarization radars. On the left of the figure are the theoretical backscatter coefficients η as computed by Rozenberg using the Marshall-Palmer drop-size distribution.[30] On the right, are measured values for 9.37 through 70 GHz obtained with a radar that transmitted and received vertical polarization signals as indicated by the V-V polarization notation.

Rozenberg classified rain as precipitation in the form of water drops with diameters in the 0.5- to 7-mm range. Drizzle was classified as precipitation not exceeding 0.25 mm/hr consisting of small droplets with diameters less than 0.5 mm. In the drizzle model of Figure 2.10, the minimum diameter of the drops was 0.1 mm and the maximum diameter was 0.5 mm. The Marshall-Palmer and Laws and Parsons distributions for the number of drops of a given size are nearly equivalent for drop-size diameters greater than 1.0 to 1.5 mm. For backscatter applications where larger drop sizes dominate, the exponential Marshall-Palmer distribution is used.[20] According to Crane, measurements of raindrop size distributions contain large variations for the same location, rain type, and rain rate. Therefore, drop-size distribution models should be regarded as representative of average, rather than individual, rain conditions.[31] Again the theory for rain backscatter coefficient adequately models the measured values.

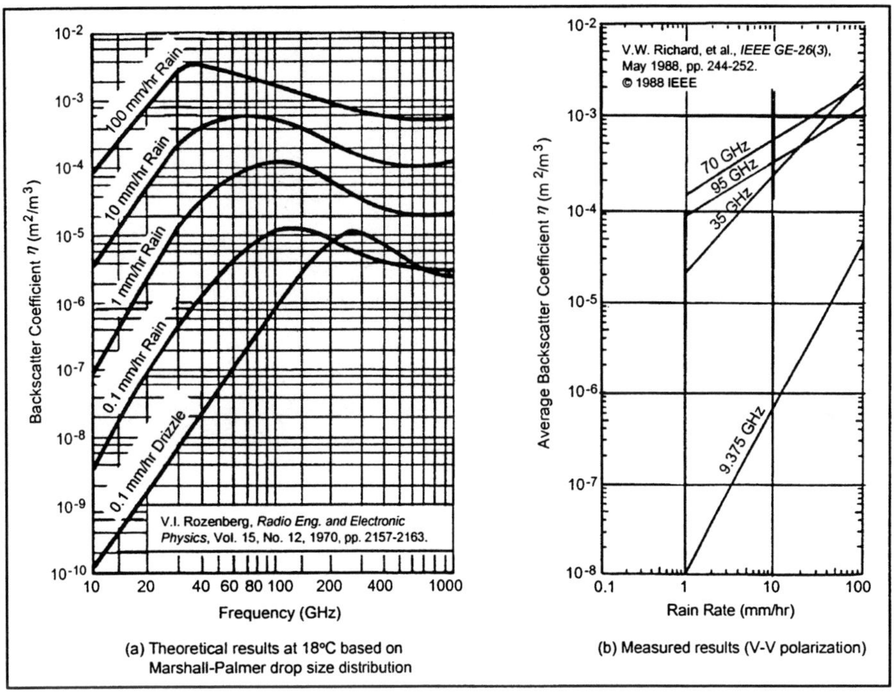

Figure 2.10: Rain backscatter coefficient as a function of frequency and rain rate

If backscatter is large at the selected frequency of operation, a potential solution is to use circular polarization. As shown in Figure 2.11, the backscatter is reduced by 20 dB at 9.375 GHz and by 18 dB at 95 GHz with this technique.[32]

2.9 Effects of operating wavelength on IR sensor performance

IR transmittance through a sea-level atmosphere is shown[33] in Figure 2.12. Unlike the attenuation data given for radar, these data show the transmittance or the percent of energy that is transmitted. The principal permanent atmospheric constituents contributing to the absorption of energy at IR wavelengths are carbon dioxide, nitrous oxide, and methane. Variable constituents include water vapor and ozone. In addition to absorption, IR energy is scattered from molecules and aerosols in the atmosphere. Wavelengths less than 2 μm experience negligible molecular scattering, while scattering from aerosols is a function of the radius of the scatterer divided by the wavelength. Aerosol-type scatterers include rain, dust, fog, and smoke.

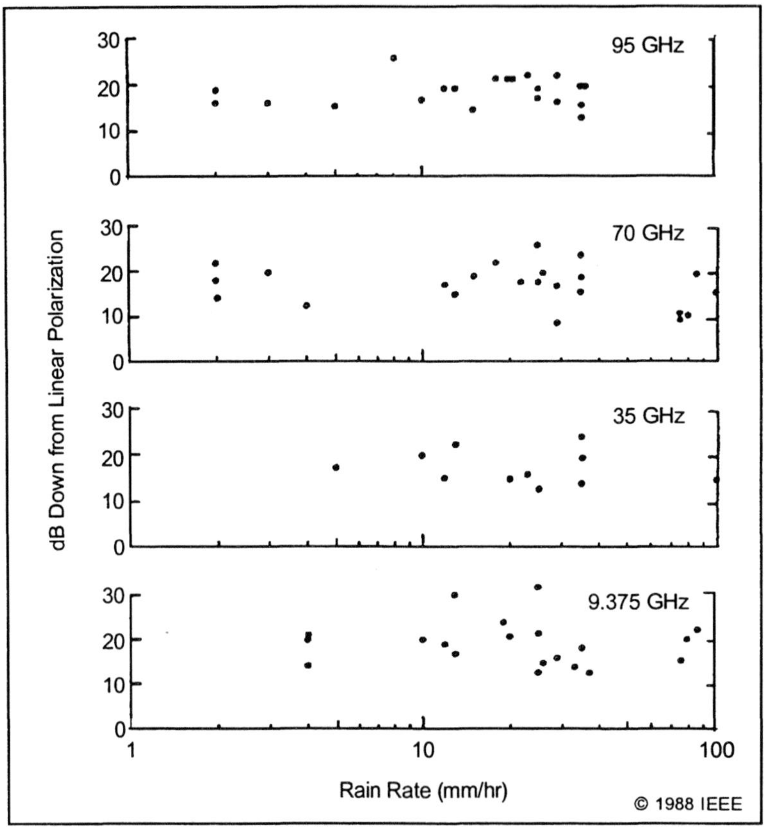

Figure 2.11: Rain backscatter coefficient reduction by circular polarization
(V.W. Richard et al., *IEEE GE-26* (3), 244-252 [May 1988])

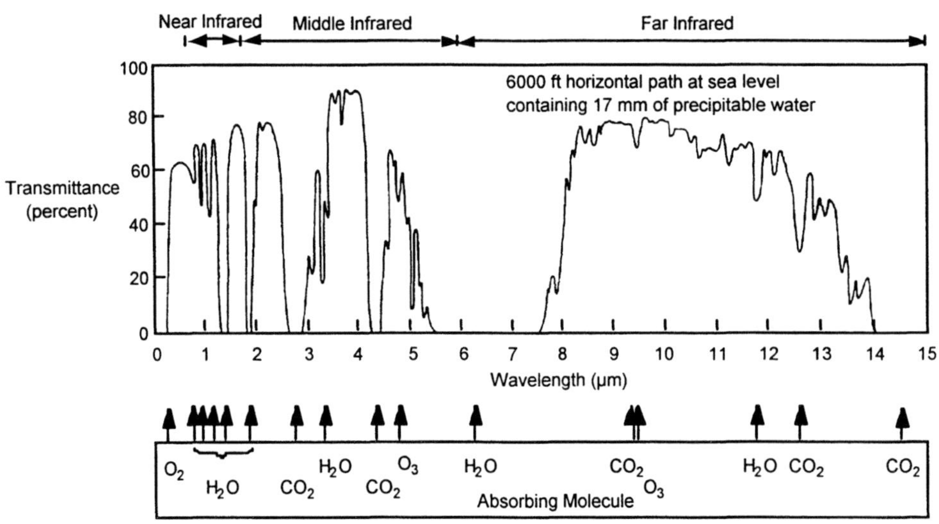

Figure 2.12: IR transmittance of the atmosphere
(R.D. Hudson, *Infrared System Engineering*, John Wiley and Sons, NY [1969])

Atmospheric transmittance $\tau_a(\lambda)$ can be modeled by the Lambert-Beer Law[34,35] as

$$\tau_a(\lambda) = \exp\,[-\gamma(\lambda)\,R] \tag{2-8}$$

where

γ = extinction coefficient or power attenuation coefficient in Np/km and

R = range or path length in km.

Nepers are the natural unit for exponents appearing in an exponential function. Multiplying the extinction coefficient in dB/km by 0.23 converts it into Np/km.

The extinction coefficient $\gamma(\lambda)$ is the sum of the absorption and scattering coefficients $k(\lambda)$ and $\sigma(\lambda)$, respectively, and can be written as

$$\gamma(\lambda) = k(\lambda) + \sigma(\lambda). \tag{2-9}$$

Absorption and scattering coefficients, in turn, are sums of molecular and aerosol components denoted by the subscripts m and a, respectively, such that

$$k(\lambda) = k_m(\lambda) + k_a(\lambda) \tag{2-10}$$

and

$$\sigma(\lambda) = \sigma_m(\lambda) + \sigma_a(\lambda). \tag{2-11}$$

The extinction coefficient is a complex function of wavelength as may be inferred from Figure 2.12. An expression for the average value of the transmission $\overline{\tau_a}$ over a wavelength band λ_1 to λ_2 is given by

$$\overline{\tau}_a = 1/(\lambda_2 - \lambda_1) \int_{\lambda_1}^{\lambda_2} \exp[-\gamma(\lambda)\,R]\,\mathrm{d}\lambda. \tag{2-12}$$

The average values of the transmission over a wavelength interval, $\overline{\tau_a}$, are generally obtained from computer-hosted programs such as LOWTRAN,[36,37,38], which spans a spectral range of 0 to 50,000 cm^{-1} (0.2 μm to infinity) with a spectral resolution of 20 cm^{-1} full width at half maximum (FWHM). LOWTRAN calculates radiance from single and multiple scattering models and path geometries corresponding to space-viewing ground-based sensors, air-to-air scenarios, surface point-to-point paths, and Earth-viewing airborne sensors. Additional information about LOWTRAN is found in Section 2.14.

2.10 Visibility metrics

Two measures of visibility are discussed in this section, the qualitative visibility observed by a human and the quantitative meteorological range.

2.10.1 Visibility

Visibility is a qualitative measure of distance. It is defined as the greatest distance at which it is just possible to see and identify with the unaided eye: (1) in the daytime, a dark object against the horizon sky and (2) at night, a known moderately intense light source.[38] Visibility measurement is subjective.

If the only visibility information available is the visibility metric observed by a human, V_{obs}, the meteorological range V can be estimated as

$$V = (1.3 \pm 0.3) V_{obs}. \tag{2.13}$$

2.10.2 Meteorological range

The quantitative meteorological range metric reported by the U.S. Weather Bureau for many localities can be used to estimate the visual range. It is based[39] on the reduction of apparent contrast produced by atmospheric attenuation at 0.55 μm. The apparent contrast C_x of a radiation source when viewed at a distance x is defined as

$$C_x = \frac{R_{sx} - R_{bx}}{R_{bx}}, \tag{2-14}$$

where R_{sx} and R_{bx} are the apparent radiance or radiant emittance of the source and background, respectively, when viewed from a distance x. The units of R_x are power per unit area. The distance at which the ratio

$$\frac{C_x}{C_0} = \frac{(R_{sx} - R_{bx}) / R_{bx}}{(R_{s0} - R_{b0})) / R_{b0}} \tag{2-15}$$

is reduced to 2 percent is defined as the meteorological range or sometimes the visual range. Equation (2-15) is usually evaluated at $\lambda = 0.55$ μm. The subscript 0 refers to the radiance measured at the source and at the background, i.e., $x = 0$. Using V to represent the meteorological range allows Eq. (2-15) to be rewritten to define the meteorological range as

$$\frac{C_{x=V}}{C_0} = 0.02. \tag{2-16}$$

If the source radiance is much greater than that of the background for any viewing distance such that $R_s >> R_b$ and the background radiance is constant such that $R_{b0} = R_{bx}$, then the meteorological range can be expressed in terms of the apparent radiance as

$$\frac{C_{x=V}}{C_0} = \frac{R_{sV}}{R_{s0}} = 0.02 \tag{2-17}$$

or

$$\ln\left(\frac{R_{sV}}{R_{s0}}\right) = -3.91. \tag{2-18}$$

The Lambert-Beer law for atmospheric transmittance $\tau_a(\lambda)$ can be used to relate the extinction coefficient (that includes both absorption and scattering effects) to the meteorological range. Consequently, the atmospheric transmittance is written as

$$\tau_a(\lambda) = \left(\frac{R_{sV}}{R_{s0}}\right) = \exp\,[-\gamma(\lambda)\,R], \tag{2-19}$$

where

γ = extinction coefficient or power attenuation coefficient in Np/km and

R = path length in km.

Upon taking the natural log of both sides of Eq. (2-19) and using Eq. (2-18), we find

$$\gamma(\lambda) = 3.91/V \text{ at } \lambda = 0.55\ \mu\text{m}. \tag{2-20}$$

Thus, the meteorological range is related to the extinction coefficient through the multiplicative constant of 3.91. This is sometimes referred to as the Koschmieder formula.[38,40]

2.11 Attenuation of IR energy by rain

Rain attenuates target-to-background contrast in IR imagery in two ways: first, by introducing an attenuation loss over the signal path to the receiver and second, by cooling the target.[41] A set of atmospheric transmission curves produced by LOWTRAN 6 for rain rates of 0, 1, 10, 30, and 100 mm/hr is shown in Figure 2.13. Here wavenumber is defined as the reciprocal of wavelength, the measurement path is 300 m, surface and dew point temperatures are both equal to 10°C, and the meteorological range is 23 km in the absence of rain. The transmittance curves in the upper portion of the figure apply to the mid- and far-IR spectral

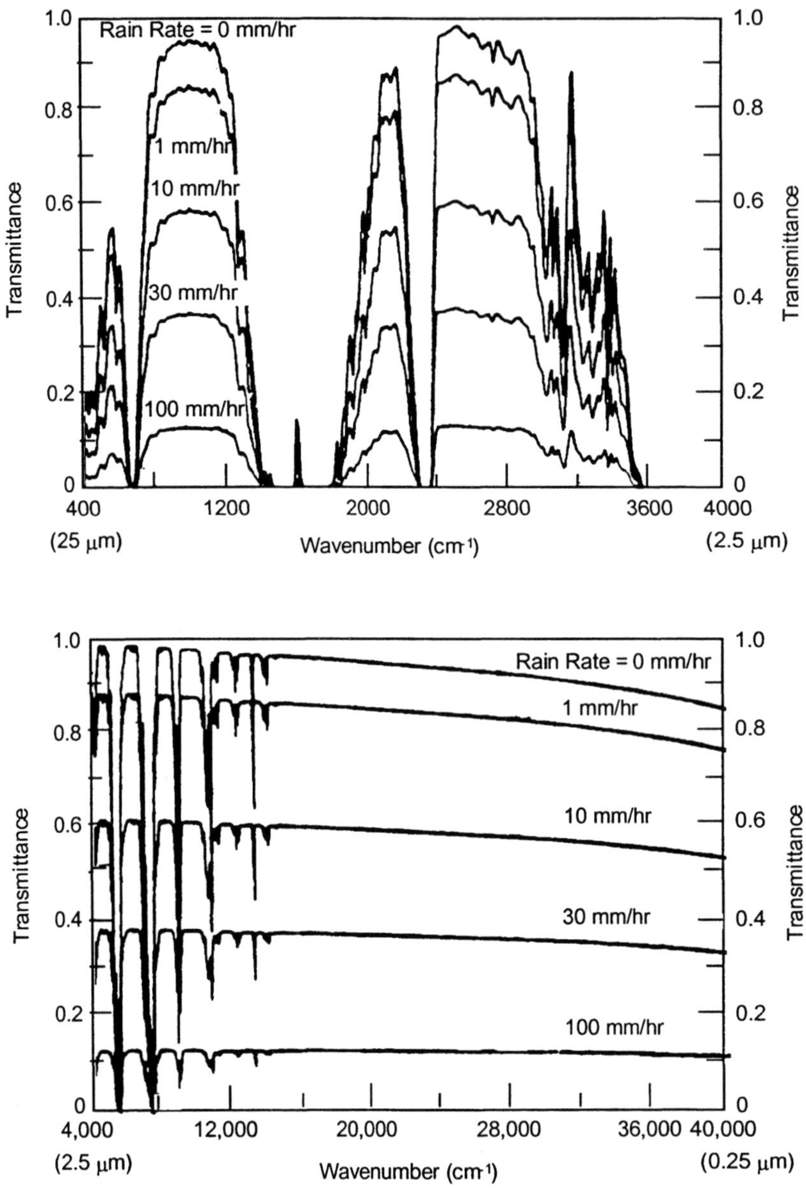

- Wavenumber = 1/wavelength
- Measurement path = 300 m
- Surface temperature = Dew point temperature = 10°C
- Meteorological range = 23 km in absence of rain

Figure 2.13: Atmospheric transmittance in rain

(F.X. Kneizys, et al., *Atmospheric Transmittance/Radiance*: *Computer Code LOWTRAN 6*, AFGL-TR-83-0187, AFGL, Hanscom AFB, MA 01731 [1983]) bands. The curves in the lower part of the figure apply to the visible and near-IR spectral bands.

2.12 Extinction coefficient values (typical)

Typical ranges for the extinction coefficients of atmospheric obscurants are shown in Table 2.5 for the visible, IR, and MMW spectral bands.[42] The extinction coefficient is expressed in units of Np/km. A qualitative correlation between visual range and extinction coefficient is presented in the lower portion of the table.

Table 2.5: Approximate ranges of extinction coefficients of atmospheric obscurants (Np/km)

	← Spectral Region →				
Atmospheric Obscurant	**Visible 0.4 to 0.7 μm**	**Mid IR 3 to 5 μm**	**Far IR 8 to 12 μm**	**MMW (35 GHz) 8.6 mm**	**MMW (95 GHz) 3.2 mm**
Gases	Very low: ≅ 0.02	Low/med: 0.25 to 0.73	Very low/med: 0.03 to 0.8	Very low: 0.02 to 0.06	Very low/low: 0.03 to 0.2
Haze	Low/med: 0.2 to 2.0	Very low/med: 0.02 to 1.0	Very low/low: 0.02 to 0.4	Very low: ≅ 0.001	Very low: ≅ 0.001
Fog	High: 2.0 to 20	Very low/med: 1.0 to 20	Med/high: 0.4 to 20	Very low/low: 0.001 to 0.1	Very low/low: 0.01 to 0.4
Rain	Low/med: 0.3 to 1.6	Low/med: 0.3 to 1.6	Low/med: 0.3 to 1.6	Very low/med: 0.05 to 1.0	Low/med: 0.3 to 2.0
Snow	Med/high: 2.0 to 12	Med/high: 2.0 to 12	Med/high: 2.0 to 12	Very low/med: 0.004 to 1.0	Very low/med: 0.03 to 1.0
Dust	Low/high: 0.2 to 4.0	Low/high: 0.2 to 4.0	Low/high: 0.2 to 4.0	Very low: 0.0005 to 0.005	Very low: 0.0005 to 0.005

Extinction Coefficient	**Descriptive Term**	**Visual Range**
< 0.1 Np/km	Very low	> 30 km, very clear
0.1 to 0.5 Np/km	Low	6 to 30 km, clear to hazy
0.5 to 2 Np/km	Medium	2 to 6 km, hazy
> 2 Np/km	High	< 2 km, foggy

2.13 Summary of advantages and disadvantages of electromagnetic sensors

Resolution, weather, day/night operation capability, clutter, and countermeasures influence the choice of particular electromagnetic sensors for object discrimination and track estimation, as shown in Table 2.6. As frequency is increased, resolution improves and designs are more compact, but degradation by the atmosphere and man-made obscurants increases and the ability to rapidly search large areas can decrease. Active sensors provide easily acquired range and velocity data, while passive sensors provide stealth operation.

2.14 Atmospheric and sensor system computer simulation models

The following sections contain descriptions of LOWTRAN, MODTRAN, FASCODE, and EOSAEL. The first three are atmospheric attenuation models. The fourth model analyzes a variety of processes that affect the performance of MMW, IR, visible, ultraviolet, and laser sensors. The material below introduces the reader to the phenomena that are treated by the models but is not meant to be a complete users manual for the computer programs.

2.14.1 LOWTRAN attenuation model

LOWTRAN 7 (rendered obsolete by MODTRAN 4) calculates atmospheric transmittance, atmospheric background radiance, single-scattered solar and lunar radiance, direct solar irradiance, and multiple-scattered solar and thermal radiance. The spectral resolution is 20 cm^{-1} (FWHM) in steps of 5 cm^{-1} from 0 to 50,000 cm^{-1}. A single parameter (absorption coefficient) is used to model molecular line absorption and molecular continuum absorption. LOWTRAN also models molecular scattering and aerosol and hydrometer absorption and scattering.

The input parameters for executing LOWTRAN 7 are contained on five main cards and thirteen optional cards. The types of information contained on each card are summarized[38] in Table 2.7.

The user specifies the geographical atmospheric model (from one of six defined by LOWTRAN 7 or from user-generated input), the altitude- and seasonal-dependent aerosol profiles, and the extinction coefficients. The six program-defined geographical atmospheric models are tropical, midlatitude summer, midlatitude winter, subarctic summer, subarctic winter, and the 1976 U.S. standard. Each atmospheric model defines the temperature, pressure, density, and atmospheric gases mixing ratio as a function of altitude. The gases modeled are water vapor, ozone, methane, carbon monoxide, and nitrous oxide. Aerosol profiles and extinction coefficients for the boundary layer (0 to 2 km), troposphere (2 to 10 km), stratosphere (10 to 30 km), and transition profiles from the stratosphere up to 100 km are provided through program-defined models and user-selected inputs. Rain rate, cloud models, wind speed, and meteoric dust

Table 2.6: Electromagnetic sensor performance for object discrimination and track estimation

Sensor	Advantages	Disadvantages
• Microwave/ millimeter-wave radar	• All weather • Lower frequencies penetrate foliage • Large search area • Day/night operation • Range and image data • Velocity data with coherent system	• Moderate resolution • Not covert • Simpler radar designs exhibit more susceptibility to corner reflector decoys and active jammers
• Microwave/ millimeter-wave radiometer	• Covert imagery • All weather • Lower frequencies penetrate foliage • Large search area • Day/night operation	• Somewhat less resolution than radar for same aperture • Large bandwidth increases susceptibility to jamming • Range data, in theory, by performing a maneuver
• Infrared imager (FLIR)	• Fine spatial and spectral resolution imagery • Covert • Day/night operation	• Affected by rain, fog, haze, dust, smoke • Poor foliage and cloud penetration • Requires cooled focal plane to maximize signal-to-noise ratio • Large search areas require scan mechanism or large detector array • Range data by performing a maneuver
• Infrared tracker (IRST)	• Hot-spot detection • Covert target tracking • Compact • Day/night operation	• Same disadvantages as infrared imager
• Laser radar	• Fine spatial and spectral resolution imagery • Range and reflectance data • Velocity and track data • Can be compact • Day/night operation	• Affected by rain, fog, haze, dust, smoke • Poor foliage penetration • Most effective when cued by another sensor to search a relatively small area
• Visible imager	• Best-resolution imager • Covert • Technology well understood	• Day operation only • Affected by clouds, rain, fog, haze, dust, smoke and any other atmospheric obscurants • No foliage penetration • No range data

extinction coefficients can be varied to tailor the aerosol profiles to the conditions under which the transmission is desired. Table 2.8 contains the characteristics of the rural, urban, maritime, tropospheric, and fog aerosol profiles that are defined by LOWTRAN.[36,37,38]

Table 2.7: LOWTRAN 7 input card information

Card	Information
1	Specifies one of six geographical-seasonal model atmospheres or a user-specified model; horizontal, vertical, or slant atmospheric path; transmittance or radiance calculation; scattering option
2	Altitude- and seasonal-dependent aerosol profiles and aerosol extinction coefficients, cloud and rain models, wind speed, altitude of surface relative to sea level
2A	Cirrus cloud altitude profile
2B	Vertical structure algorithm of aerosol extinction and relative humidity for low visibility or low ceiling conditions as occur with: (1) cloud/fog at the surface, (2) hazy/light fog, (2') clear/hazy, (3) radiation fog or inversion layer, (4) no cloud ceiling or inversion layer
2C	Additional data for user-defined atmospheric model (if selected on Card 1)
2C1	Additional data for user-defined atmospheric model (if selected on Card 1)
2C2	Additional data for user-defined atmospheric model (if selected on Card 1)
2C3	Additional data for cloud, fog, and rain user-defined atmospheric model (if selected on Card 1)
2D	User-defined attenuation coefficients for any or all four of the aerosol altitude regions (boundary layer, troposphere, stratosphere, above stratosphere to 100 km)
2D1	Conversion factor from equivalent liquid water content (g/m^3) to extinction coefficient (Np/km)
2D2	User-defined aerosol or cloud extinction coefficients, absorption coefficients, and asymmetry parameter
3	Geometrical path parameters
3A1	Solar/lunar scattered radiation
3A2	Additional parameters for solar/lunar scattered radiation
3B1	User-defined phase functions
3B2	Additional parameters for user-defined phase functions
4	Spectral range and calculation increment (frequency step size in cm^{-1})
5	Recycle parameter to iterate the calculations through the program so that a series of problems can be run with one submission of LOWTRAN

Table 2.8: LOWTRAN aerosol profiles

Aerosol Model	Representative Region	Constituent	Default Visibility*
Rural (0 to 2 km altitude)	Continental areas not directly influenced by urban/industrial aerosol sources	Atmospheric gases and surface dust particles	23 or 5 km
Urban (0 to 2 km altitude)	Modifies rural background by adding aerosols from combustion products and industrial sources	20%/80% mixture of carbonaceous aerosols to rural type aerosols, respectively	5 km
Maritime (0 to 2 km altitude)	Aerosols of oceanic origin	Sea salt particles	User selected or 23 km
Tropospheric (2 to 10 km altitude)	Troposphere with extremely clear conditions and uniform aerosol properties	Rural model constituents without large particles	50 km
Fog 1 (0 to 2 km altitude)	Advection fog	Water droplets	0.2 km
Fog 2 (0 to 2 km altitude)	Radiation fog	Water droplets	0.5 km

* Visibility refers to the surface meteorological range.

2.14.2 FASCODE and MODTRAN attenuation models

Other models available to assess the effects of weather on sensor systems are FASCODE and MODTRAN 4. These are supported by the U.S. Air Force Geophysics Laboratory at Hanscom Air Force Base, Bedford, Massachusetts 01731.[43-50] FASCODE is available from the Geophysics Laboratory via an e-mail request. MODTRAN 4 is available by signing a non-disclosure agreement with the U.S. Air Force and payment of a licensing fee for downloading privileges and support. A CD version of MODTRAN 4 can be obtained from Ontar Corporation once the non-disclosure agreement is signed and fees are paid.[2]

[2]. The contact personnel at the Geophysics Laboratory are Gail Anderson: GAnderson@plh.af.mil and Jim Chetwynd: Chetwynd@plh.af.mil. The latest description of MODTRAN 4 and licensing information is available at http://www-vsbm.plh.af.mil/soft/modtran4.html. Ontar Corporation's web site is Ontar.com.

FASCODE models very high altitude (>70 km) and very narrow spectral bands that are applicable to laser-line resolution. FASCODE is useful for extinction dominated by molecular absorption, improving upon the resolution offered by LOWTRAN in this region. MODTRAN was written for moderate resolution calculations that do not require FASCODE. Originally an enhanced version of LOWTRAN 7, MODTRAN contains six additional routines that increase the 20 cm^{-1} spectral resolution found in LOWTRAN to 2 cm^{-1} (FWHM) resolution for wavelengths less than 380 nm. Ultraviolet frequencies retain the 20 cm^{-1} spectral resolution, although future resolution improvements are planned for this region as well.[44] MODTRAN models the molecular absorption by atmospheric molecules as a function of temperature and pressure and provides new capabilities for calculating three absorption band parameters for thirteen molecular species (water vapor, carbon dioxide, ozone, nitrous oxide, carbon monoxide, methane, oxygen, nitric oxide, sulfur dioxide, nitrogen dioxide, ammonia, nitric acid, and oxygen-hydrogen). The absorption band parameters in MODTRAN are temperature dependent and include an absorption coefficient, a line density parameter, and an average linewidth. LOWTRAN 7 uses only the absorption coefficient and molecular density scaling functions to define the absorption band. MODTRAN 4 offers an improved multiple scattering model for more accurate transmittance and radiance calculations that facilitate the analysis of hyperspectral imaging data.[51] Sets of bi-directional radiance distribution functions (BRDFs) have been provided to support surface scattering distributions other than Lambertian.

MODTRAN has been successfully compiled and executed under UNIX using a FORTRAN 77 compiler and in IBM-compatible personal computers using a binary source code with a Lahey F77L32 compiler. All the usual LOWTRAN options such as aerosol profiles, path selection, multiple scattering models, and user-specified inputs have been maintained in MODTRAN.

The input data sequence for MODTRAN is identical to LOWTRAN 7 except for one modification to Card 1 and two modifications to Card 4. A logical parameter MODTRN has been added to the front end of Card 1 to act as a switch. When set to F (false), it causes LOWTRAN 7 to execute. When set to T (true), it activates MODTRAN. The input to Card 4 has been changed to integer format and a resolution parameter IFWHM added as the last entry on the card. IFWHM is only read if MODTRN is true, specifying the full width at half maximum of an internal triangular slit function that determines the 2 cm^{-1} spectral resolution of the program.

2.14.3 EOSAEL sensor performance model

One of the more comprehensive models for analyzing a variety of physical processes that affect the performance of MMW and IR sensors, as well as those that operate in the visible, ultraviolet, and on 53 laser lines, is EOSAEL (Electro-

Optical Systems Atmospheric Effects Library).[52-53] The aspects of electromagnetic energy propagation and defense application scenarios addressed by the model are:

- Spectral transmission and contrast transmission;
- Multiple scattering;
- Sensor performance;
- Transport and diffusion;
- Turbulence effects on imaging;
- High-energy laser propagation;
- Radiative transfer;
- Thermal contrast;
- Generation of battlefield obscurants;
- Climatology for 47 nonoverlapping climatic regions.

EOSAEL 92 (the latest version) software contains 27 modules that are divided into nine categories: gaseous transmission, natural aerosols, battlefield aerosols, radiative transfer, laser propagation, waves and acoustic, target acquisition, decision aides, and support. The modules are more engineering oriented than based on first principles. The EOSAEL development philosophy was to include modules that give reasonably accurate results, while minimizing computer time, for conditions that may be expected on a battlefield. The modules can be executed under the control of an executive routine or in a stand-alone mode. Versions are available in FORTRAN 77 that allow the program to run on a variety of computers.

The modules and functions contained in EOSAEL 87, Version 2 (March 1990) are listed in Table 2.9. Three modules of particular interest are the previously discussed LOWTRAN and NMMW (both found in the gases category) and TARGAC. NMMW models the effects of atmospheric gases on MMW sensors. TARGAC, in the target acquisition and system performance category, is built into the FLIR performance model developed by the U.S. Army Center for Night Vision and Electro-Optics (CNVEO).[54-56] The FLIR performance model describes the relation of the target-to-background contrast temperature to the sensor resolution and the range between the sensor and target at which the target can be detected, classified, or identified. Ontar Corporation supplies a demonstration FORTRAN source code version of EOSAEL 87, Version 2 (March 1990) on a CD at no charge. The newer EOSAEL 92 can be procured from Ontar as well.

2.15 Summary

The attributes of active and passive sensors in the microwave, millimeter-wave, and infrared portions of the electromagnetic spectrum have been enumerated to illustrate the advantages they bring to a high-performance, multisensor suite in defense and civilian applications. The selection of MMW and IR sensor operating frequencies has an impact on resolution, hardware availability and specifications, and compatibility with the expected signatures from the objects of interest and the backgrounds in which the sensors operate. In civilian applications, the longer wavelength microwave and millimeter-wave sensors penetrate clouds and provide data used in weather forecasting, pollution and Earth resource management, and land-use monitoring. Multispectral IR imagery provides information about land cover and geological features, cloud cover, river expansion from floods, and changes in the ocean ecosystem. The relatively good performance of the active mode microwave and MMW sensors in inclement weather and various countermeasures can be used to complement an IR sensor to provide reliable target detection, range, and tracking for military applications. The higher resolution IR sensors provide imagery for classifying potential military targets and improving the selection of a missile impact point.

Measured data and models were presented for calculating atmospheric absorption and backscatter of MMW and IR energy in clear weather, rain, and fog. Attenuation of MMW and IR energy may be modeled using an extinction coefficient that contains terms to account for absorption and scattering. The modeled data generally agree with measured data and, therefore, can be used to predict sensor performance when actual absorption and backscatter measurements are not available. Some of the models only address atmospheric effects, while others, such as EOSAEL, model more complex problems and scenarios.

Table 2.9: EOSAEL 87 modules and their functions

Category	Module	Valid Range	Function
Gaseous Transmission	LOWTRAN	0.25 to 28.5 μm	Calculates atmospheric transmittance, radiance, and contrast due to specific molecules at 20 inverse cm spectral resolution on a linear wave-number scale
	LZTRAN	Visible to far IR (0.5 to 11.0 μm)	Calculates transmission through atmospheric gases at specific laser frequencies for slant or horizontal paths
	NMMW	10 to 1000 GHz (0.3 to 30.0 mm)	Calculates transmission, backscatter, and refractivity due to gaseous absorption, fog, rain, and snow
Natural Aerosols	XSCALE	0.2 to 12.5 μm	Calculates fog and haze transmission for horizontal or slant paths and rain and snow transmission for horizontal paths
	CLIMAT	Not applicable	Provides values of meteorological parameters for select European, Mideastern, Korean, Alaskan, Scandinavian, Central American, Indian, SE Asian, South American, and Mexican locales
	CLTRAN	0.2 to 2.0 μm 3.0 to 5.0 μm 8.0 to 12.0 μm	Calculates slant path transmission through six cloud types
	COPTER	0.4 to 0.7 μm 3.0 to 5.0 μm 8.0 to 12.0 μm 0.3 to 30.0 mm	Calculates effects of loose snow or dust lofted by helicopter downwash
Battlefield Aerosols	COMBIC	0.4 to 1.2 μm 3.0 to 5.0 μm 8.0 to 12.0 μm and 94 GHz (3 mm)	Calculates size, path length, concentration, and transmission through various smokes and artillery or vehicular dirt and dust particles
	SABRE	Not applicable	Calculates effects of terrain on wind flow and smoke screens
	KWIK	Not applicable	Provides a placement and number of smoke munitions necessary to produce a screen that reduces the probability of target detection to a given level
	GRNADE	0.4 to 1.2 μm 3.0 to 5.0 μm 8.0 to 12.0 μm and 94 GHz (3 mm)	Models obscuration produced by tube-launched grenades used in self-screening applications
	FITTE	0.4 to 12.0 μm	Calculates dimensions of and transmittance through plumes from burning vegetation and vehicles

Table 2.9: EOSAEL 87 modules and their functions (continued)

Category	Module	Valid Range	Function
Battlefield Aerosols (continued)	MPLUME	Not applicable	Calculates performance degradation of target designation systems by missile smoke plumes
Radiative Transfer	FCLOUD	Any wavelength included in the phase function file PFNDAT*	Calculates beam transmittance, path radiance, and contrast transmittance through a homogeneous ellipsoidal cloud
	OVRCST	Any wavelength	Calculates beam transmittance, path radiance, and contrast transmittance along an arbitrary line of sight under an overcast sky
	MSCAT	Any wavelength included in the phase function file PFNDAT	Calculates multiple scattering of a laser beam into a detector from an ellipsoidal aerosol cloud
	ASCAT	Any wavelength included in the phase function file PFNDAT	Calculates effects of multiple scattering for transmissometer and lidar configurations
	ILUMA	Photopic	Predicts natural illumination under realistic atmospheric conditions
	FASCAT	0.55 and 1.06 μm	Determines path radiance and contrast effects
	GSCAT	Visible	Determines path radiance effects
	LASS	Visible	Determines the effectiveness of smoke screens deployed against large fixed and semifixed installations
	REFRAC	≈0.4 to ≈20.0 μm	Calculates the amount of curvature a ray of light experiences as it passes over a complex terrain surface
Laser Propagation	IMTURB	<14 μm	Predicts the effects of laser propagation through optical turbulence in the weak fluctuation regime
	NOVAE	<14 μm	Calculates linear and nonlinear effects on high-energy laser beam propagation from clear air, smokes, and aerosols
Target Acquisition and System Performance	TARGAC	Visible to mid-IR	Evaluates the combined atmospheric and system effects to determine the range for target detection and classification
	RADAR	0.1 to 350 GHz	Calculates millimeter-wave system performance

* PFNDAT contains phase functions, extinction and scattering coefficients, and the single-scattering albedo for 38 natural and man-made aerosols at 16 wavelengths ranging from 0.55 to 12.0 μm. The single-scattering albedo is the ratio of the scattering coefficient to the extinction coefficient.

References

1. J.T. Houghton and F.W. Taylor, "Remote sounding from artificial satellites and space probes of the atmospheres of the Earth and the planets," *Rep. Prog. Phys*., 36, 827-919 (1973).
2. E.G. Njoku, "Passive microwave remote sensing of the Earth from space – a review," *Proc. IEEE*, 70(3), 728-750 (1982).
3. P.W. Rosenkranz et al., "Microwave radiometric measurements of atmospheric temperature and water from an aircraft," *J. Geophys. Res.*, 77(30), 5833-5844 (1972).
4. F.T. Ulaby, R.K. Moore, and A.K. Fung, *Microwave Remote Sensing Active and Passive,* Vol. III: *From Theory to Applications*, Artech House, Norwood, MA (1986).
5. N.C. Grody, "Surface identification using satellite microwave radiometers," *IEEE Trans. Geosci. Remote Sensing*, GE-26(6), 850-859 (1988).
6. T. Bellerby, M. Taberner, A. Wilmshurst, M. Beaumont, E. Barrett, J. Scott, and C. Durbin, "Retrieval of land and sea brightness temperatures from mixed coastal pixels in passive microwave data," *IEEE Trans. Geosci. Remote Sensing*, GE-36(6), 1844-1851 (Nov. 1998).
7. P.S. Chang and L. Li, "Ocean surface wind speed and direction retrievals from the SSM/I," *IEEE Trans. Geosci. Remote Sensing*, GE-36(6), 1866-1871 (Nov. 1998).
8. R.V. Engeset and D.J. Weydahl, "Analysis of glaciers and geomorphology on Svalbard using multitemporal ERS-1 SAR images," *IEEE Trans. Geosci. Remote Sensing*, GE-36(6), 1879-1887 (Nov. 1998).
9. N.J. Willis, *Bistatic Radar*, Artech House, Norwood, MA (1991).
10. A.O. Aboutalib and T.K. Luu, "An efficient target extraction technique for laser radar imagery," *Digital Signal Processing, Association, and Tracking of Point Source, Small, and Cluster Targets*, *Proc. SPIE* 1096 (1989).
11. J.M Lloyd, *Thermal Imaging Systems*, Plenum Press, New York, NY (1982).
12. L.A. Klein, *Millimeter-wave and Infrared Multisensor Design and Signal Processing*, Artech House, Norwood, MA (1997).
13. D.L. Foiani and R.H. Pearce, "Combined radar and radiometer at millimeter wavelengths," *Symposium on Submillimeter Waves*, Polytechnic Institute of Brooklyn, Brooklyn, NY (March 1970).
14. V.G. Plank, R.O. Berthel, and B.A. Main, "Snow characterization measurements and E/O correlation obtained during Snow-One-A and Snow-One-B," *Proc. SPIE* 414, 97-102, (1983).
15. G.W. Aitken, Ed., *Snow-One-A data report*, Special Report 82-8, AD B068569L, U.S. Army Cold Regions Research and Engineering Laboratory/ CRREL-RG, Hanover, NH (May 1982).
16. M.A. Seagraves and J.F. Ebersole, "Visible and infrared transmission through snow," *Opt. Eng.*, 22(1), 90-93 (Jan.-Feb. 1983).
17. V.J. Falcone Jr., L.W. Abreu, and E.P. Shettle, *Atmospheric attenuation of millimeter and submillimeter waves: models and computer code*, AFGL-TR-

79-0253 (AD A084485), Air Force Geophysics Laboratory, Hanscom AFB, MA (Oct. 1979).

18. F.T. Ulaby, R.K. Moore, and A.K. Fung, *Microwave Remote Sensing: Active and Passive,* Vol. I: *Microwave Remote Sensing Fundamentals and Radiometry*, Artech House, Norwood, MA (1981).
19. R.K. Parker and R.H. Abrams, Jr., "Radio frequency vacuum electronics: a resurgent technology for tomorrow," *Millimeter-Wave Technology IV and Radio Frequency Power Sources, Proc. SPIE* 791, 2-12 (1987).
20. R.K. Crane, *Microwave scattering parameters for New England rain*, Technical Report 426, Lincoln Laboratory, MIT (Oct. 3, 1966).
21. J. Nemarich, R.J. Wellman, and J. Lacombe, "Backscatter and attenuation by falling snow and rain at 96, 140, and 225 GHz," *IEEE Trans. Geosci. Remote Sensing*, GE-26(3), 319-329 (1988).
22. H. Goldstein, "Attenuation by condensed water," Chapter 8 in *Propagation of Short Radio Waves*, Editor: D.E. Kerr, McGraw-Hill, New York, NY (1951).
23. D. Atlas, "Advances in radar meteorology," in *Advances in Geophysics,* Vol. 10, 317-478, Academic Press, New York, NY (1964).
24. A.J. Bogush, Jr., *Radar and the Atmosphere*, Artech House, Norwood, MA (1989).
25. R.K. Crane, "Prediction of attenuation by rain," *IEEE Trans. Comm.*, COM-28(9), 1717-1733 (Sept. 1980).
26. V.J. Falcone, Jr. and L.W. Abreu, "Atmospheric attenuation of millimeter and submillimeter waves," *IEEE EASCON-79 Conference Record*, Vol. 1, pp. 36-41 and *Millimeter Wave Radar*, S.L. Johnston, Editor, Artech House, Norwood, MA (1980).
27. V.W. Richard, J.E. Kammerer, and R.G. Reitz, *140 GHz Attenuation and Optical Visibility Measurements of Fog, Rain, and Snow*, ARBRL-MR-2800 (1977).
28. K.A. Richer, "Environmental effects on radar and radiometric systems at millimeter wavelengths," A72-15610, *Proc. Symposium on Submillimeter Waves*, Polytechnic Institute of Brooklyn, Brooklyn, NY, 533-543 (Mar. 31-Apr. 2, 1970).
29. H.B. Wallace, "Millimeter-wave propagation measurements at the Ballistic Research Laboratory," *IEEE Trans. Geosci. Remote Sensing*, GE-26(3), 253-258 (1988).
30. V.I. Rozenberg, "Radar characteristics of rain in submillimeter range," *Radio Eng. and Electron. Phys.*, 15(12), 2157-2163 (1970).
31. R.K. Crane, "Propagation phenomena affecting satellite communication systems operating in the centimeter and millimeter wavelength bands," *Proc. IEEE*, 59, 173-188 (1971).
32. V.W. Richard, J.E. Kammerer, and H.B. Wallace, "Rain backscatter measurements at millimeter wavelengths," *IEEE Trans. Geosci. Remote Sensing*, GE-26(3), 244-252 (1988).

33. R.D. Hudson, *Infrared System Engineering*, John Wiley and Sons, New York, NY (1969).
34. H. Weichel, *Laser Beam Propagation in the Atmosphere,* Vol. TT3, SPIE Optical Engineering Press, Bellingham, WA (1990).
35. A.J. LaRocca, "Methods of calculating atmospheric transmittance and radiance in the infrared," *Proc. IEEE*, 63(1), 75-94 (Jan. 1975).
36. F.X. Kneizys, E.P. Shettle, W.O. Gallery, J.H. Chetwynd Jr., L.W. Abreu, J.E.A. Selby, R.W. Fenn, and R.A. McClatchey, *Atmospheric Transmittance/ Radiance: Computer Code LOWTRAN 5*, AFGL-TR-80-0067, AFGL, Hanscom AFB, MA 01731 (1980).
37. F.X. Kneizys, E.P. Shettle, W.O. Gallery, J.H. Chetwynd Jr., L.W. Abreu, J.E.A. Selby, S.A. Clough, and R.W. Fenn, *Atmospheric Transmittance/ Radiance: Computer Code LOWTRAN 6*, AFGL-TR-83-0187, AFGL, Hanscom AFB, MA 01731 (1983).
38. F.X. Kneizys, E.P. Shettle, L.W. Abreu, J.H. Chetwynd Jr., G.P. Anderson, W.O. Gallery, J.E.A. Selby, and S.A. Clough, *Users Guide to LOWTRAN 7*, AFGL-TR-88-0177, AFGL, Hanscom AFB, MA 01731 (1988).
39. P.W. Kruse, L.D. McGlauchlin, and R.B. McQuistan, *Elements of Infrared Technology*, John Wiley and Sons (1962).
40. H.A. Brown and B.A. Kunkel, "Water vapor, precipitation, clouds, and fog," Section 4 of Chapter 16 in *Handbook of Geophysics and the Space Environment*, Editor: A.S. Jursa, AFGL-TR-85-0315 (AD A167000), USAF Geophysics Laboratory, Hanscom AFB, MA 01731 (1985).
41. W.R. Watkins, F.T. Kantrowitz, and S.B. Crow, "Optical, infrared, and millimeter wave propagation engineering," *Proc. SPIE* 926, 69-84 (1988).
42. *Smoke and Natural Aerosol Parameters* (*SNAP*) *Manual*, Joint Technical Coordinating Group for Munitions Effectiveness, Smoke and Aerosol Working Group, Report 61, JTCG/ME-85-2 (Apr. 26, 1985).
43. R.W. Fenn, S.A. Clough, W.O. Gallery, R.E. Good, F.X. Kneizys, J.D. Mill, L.S. Rothman, E.P. Shettle, and F.E. Volz, "Optical and Infrared Properties of the Atmosphere," Chapter 18 in *Handbook of Geophysics and the Space Environment*, Editor: A.S. Jursa, AFGL-TR-85-0315 (AD A167000), USAF Geophysics Laboratory, Hanscom AFB, MA 01731 (1985).
44. A. Berk, L.S. Bernstein, and D.C. Robertson, *MODTRAN: A Moderate Resolution Model for LOWTRAN 7*, GL-TR-89-0122, USAF Geophysics Laboratory, Hanscom AFB, MA 01731 (1989).
45. L.S. Rothman, et al., "The HITRAN Molecular Database: Editions of 1991 and 1992," *J. Quant. Spectrosc. Radiat. Transfer*, 28, 469 (1992).
46. G.P. Anderson, J.H. Chetwynd, J.M. Theriault, P. Acharya, A. Berk, D.C. Robertson, F.X. Kneizys, M.L. Hoke, L.W. Abreu, and E.P. Shettle, "MODTRAN 2: suitability for remote sensing," *Atmospheric Propagation and Remote Sensing II, Proc. SPIE* 1968, 514-525 (1993).
47. P.K. Acharya, D.C. Robertson, and A Berk, *Upgraded Line-of-Sight Geometry Package and Band Model Parameters for MODTRAN*, Phillips

Laboratory Technical Report PL-TR-93-2127, Geophysics Directorate, Hanscom AFB, MA 01731 (1993).
48. A. Berk, *Upgrades to the MODTRAN Layer Cloud/Rain Models*, Report SSI-SR-56, Spectral Sciences, Inc., Burlington, MA (1995).
49. J.H. Chetwynd, J. Wang, and G.P. Anderson,, "FASCODE: an update and applications in atmospheric remote sensing," *Proc. SPIE* 2266, 613-623 (1994).
50. J. Wang, G.P. Anderson, H.E. Revercomb, and R.O. Knutseon, "Validation of FASCODE 3 and MODTRAN 3: comparison of model calculations with ground-based and airborne interferometer observations under clear sky conditions," *Appl. Op.*, 35(30), 6028-6040 (Oct. 1996).
51. L.S. Bernstein, A. Berk, D.C. Robertson, P.K. Acharya, G.P. Anderson, and J.H. Chetwynd, "Addition of a Correlated-k Capability to MODTRAN," *Proc. 19th Annual Conf. on Atmospheric Transmission Models*, Phillips Laboratory, Geophysics Directorate, Hanscom AFB, MA 01731 (Jun. 1996).
52. R.C. Shirkey, "Effects of atmospheric and man-made obscurants on visual contrast," *Atmospheric Effects on Systems Performance, Proc. SPIE* 305, 37-44 (1981).
53. R.C. Shirkey, L.D. Duncan, and F.E. Niles, *EOSAEL 87, Vol. 1, Executive Summary*, Rpt. TR-0221-1, U.S. Army Laboratory Command, Atmospheric Sciences Laboratory, White Sands Missile Range, NM 88002-5501 (Oct. 1987).
54. Night Vision Laboratory Static Performance Model for Thermal Viewing Systems, ECOM Report 7043, AD A011212 (Apr. 1975).
55. L. Scott and L. Conduff, "CNVEO advanced FLIR systems performance model," *Proc. SPIE* 1309, Paper 17 (Apr. 1990).
56. H.V. Kennedy, "Modeling second generation thermal imaging sensors," *Proc. SPIE* 1309, Paper 1 (Apr. 1990).

Chapter 3

Data Fusion Algorithms and Architectures

Sensor and data fusion are used in applications ranging from Earth resource monitoring, weather forecasting, and vehicular traffic control to military target classification and tracking. Data fusion and its objectives are discussed in this chapter based on a model developed for the U.S. Department of Defense. The model divides data fusion into low-level and high-level processes. The low-level processes support preprocessing of data; target detection, classification, and identification; and target tracking. High-level processes support situation and threat assessment and fusion process refinement. Various classes of algorithms have been developed to implement target detection, classification, and track estimation fusion. In addition, several types of data fusion architectures exist for combining sensor data to support the requirements of the data fusion model. The architectures are differentiated by the amount of processing applied to the sensor data and the resolution of the data that are combined. The chapter concludes by addressing several concerns associated with the registration of multisensor data. These issues encompass dissimilar sensor footprint sizes, signal generation phenomena, and uncertainty in the location of the sensors.

3.1 Definition of data fusion

In an effort to encourage the use of sensor and data fusion to enhance target discrimination, target tracking, and situation and threat assessment in real time with affordable, survivable, and maintainable systems, the Assistant Secretary of Defense for C^3I (Command, Control, Communications, and Integration) empowered the Joint Directors of Laboratories Data Fusion Subpanel (JDL DFS) to codify data fusion terminology and improve the efficiency of data fusion programs through the exchange of technical information.[1] Acting on this directive, the Office of Naval Technology (ONT) chartered a group, the Data Fusion Development Strategy (DFDS) Panel, to devise a plan for guiding future ONT investment in data fusion.[2] The results of their activity form the basis for the definition, objectives, and functional description of data fusion presented here. This definition has been enhanced by Waltz and Llinas, who added detection to the functions performed by data fusion and replaced the estimation of position by the estimation of state "to include the broader concept of kinematic state (e.g., higher order derivatives such as velocity) as well as other states of behavior (e.g., electronic state, fuel state)."[3] The resulting definition of data fusion is:

A multilevel, multifaceted process dealing with the automatic detection, association, correlation, estimation, and combination of data and information from single and multiple sources.

The goals of data fusion are realized through a four-level hierarchy of processing as shown in Figure 3.1. A fifth level, Level 0, is sometimes assigned to the source preprocessing function. The processing that occurs at the different data fusion levels is described as follows:

- Level 0 processing: satisfies process estimation and processor computational and scheduling requirements by normalizing, formatting, ordering, batching, and compressing input data;

- Level 1 processing: achieves refined position and identity estimates by fusing individual sensor position and identity estimates;

- Level 2 processing: assists in complete and timely hostile or friendly military situation assessment;

- Level 3 processing: assists in complete and timely force threat assessment;

- Level 4 processing: achieves improved results by continuously refining estimates and assessments, evaluating the need for additional sources, or modifying the process itself.

Data gathered from all appropriate sources, including real-time sensor information, intelligence, maps, weather reports, friendly or hostile status of targets, threat level of targets (e.g., immediate, imminent, and potential), prediction of probable intent and strategies of the threatening targets, and information from other databases, are input to the fusion domain as illustrated on the left of Figure 3.1. The data may be subject to preprocessing or pass directly into one of the other fusion levels. A significant amount of information from external databases is usually needed to support the Level 2 and 3 fusion processes. Interrelationships among Level 1 through 3 fusion processes are illustrated in Figure 3.2. In some applications such as missile tracking, target discrimination and target tracking occur simultaneously rather than in separate paths as displayed in Figure 3.2.

3.2 Level 1 processing

Level 1 processing is the low-level processing resulting in target track estimation and target discrimination.[4] There is actually a hierarchy of discrimination that, from lowest to highest, encompasses detection, orientation, classification (also called recognition in the older literature), and identification. The interpretation

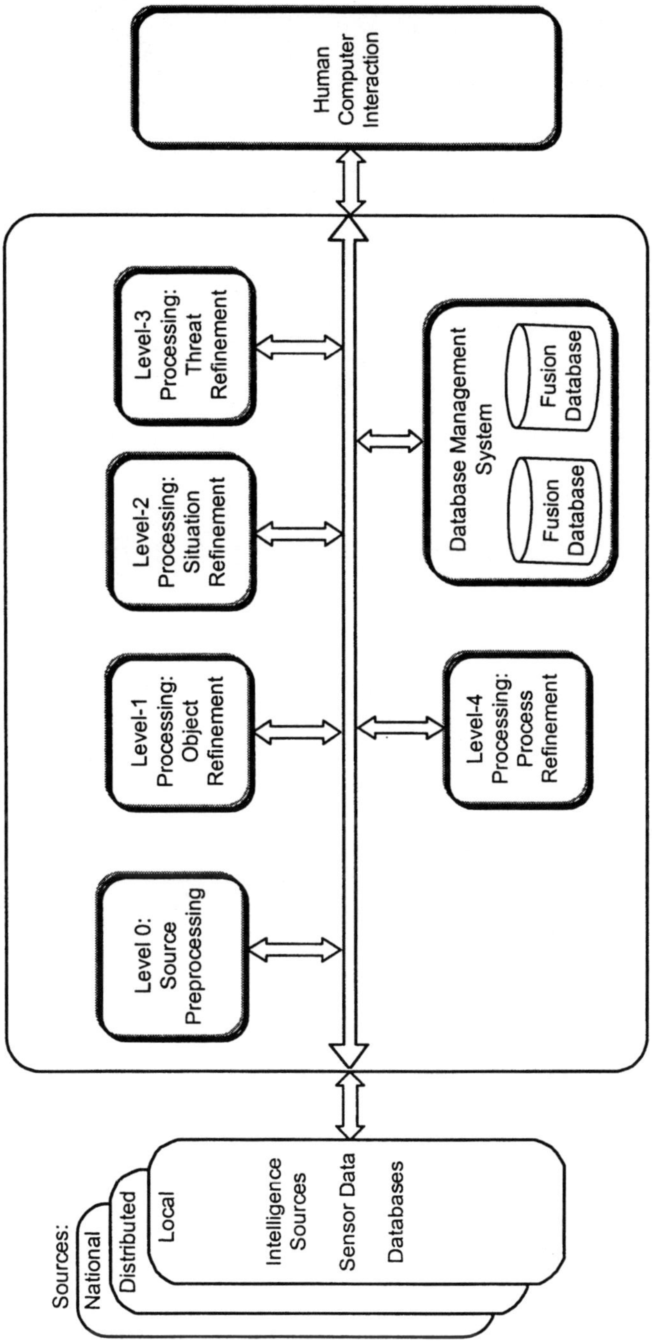

Figure 3.1: Data fusion model showing processing levels 0, 1, 2, 3, and 4

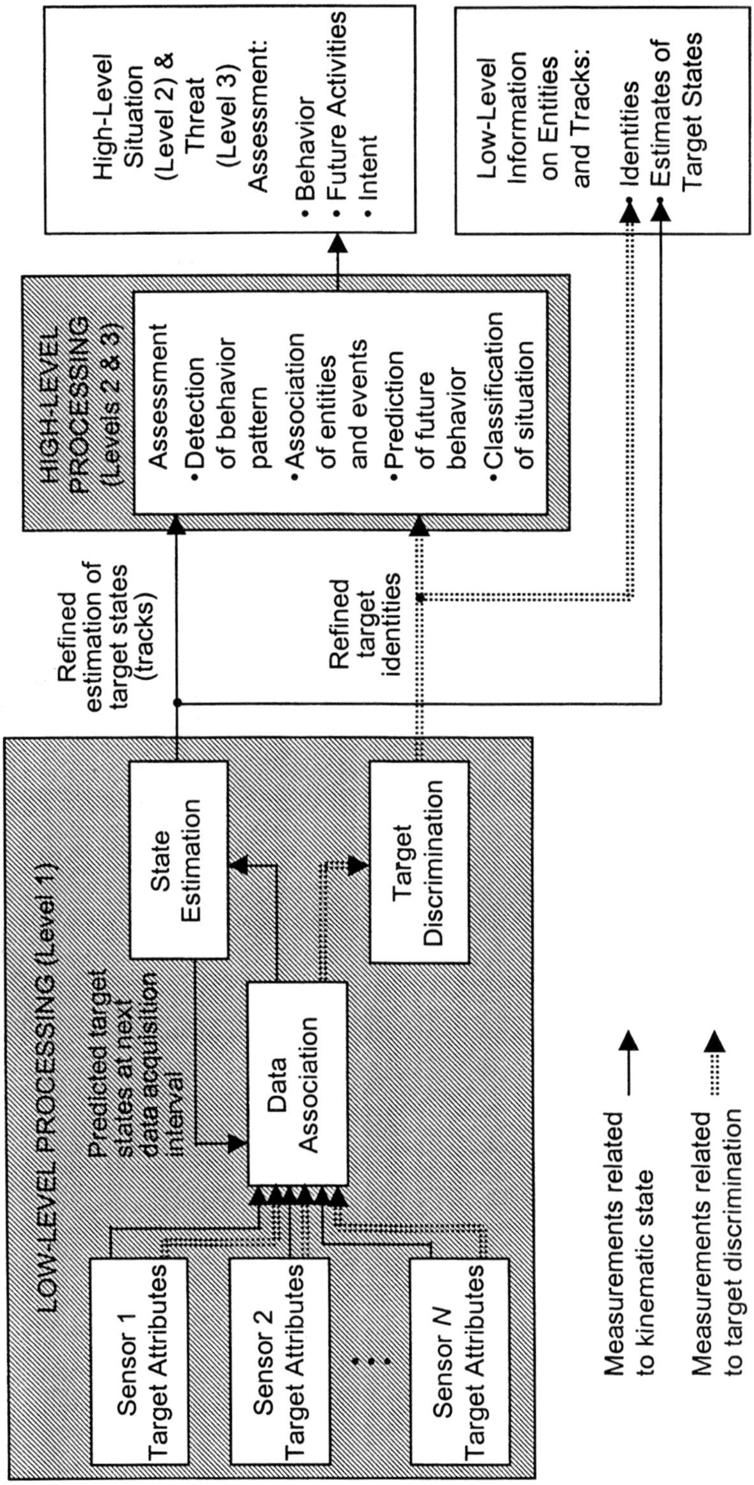

Figure 3.2: Data fusion processing levels 1, 2, and 3
(Adapted from E. Waltz and J. Llinas, *Multisensor Data Fusion*, Artech House, Norwood, MA [1990])

of these terms is shown[5,6,7] in Table 3.1. The ability to achieve a given level of discrimination depends on the resolution of the sensor and the signal-to-noise ratio at the input to the sensor. These parameters may be traded off against each other to achieve detection, classification, and identification requirements.[6,7,8]

Table 3.1: Object discrimination categories

Category	Interpretation
Detection	Object is present
Orientation	Object is discerned as approximately symmetric or asymmetric and its orientation is determined
Classification	Class to which object belongs is discerned (e.g., building, truck, tank, man, trees, field)
Identification	Object is described to limit of an observer's knowledge (e.g., motel, pickup truck, T-62 tank, M-105 howitzer, soldier)

Sensor outputs are combined through data association processes to produce the desired object or target discrimination level and target position estimate and track. The fusion algorithm used for the detection and classification process need not be the same as that used for track estimation and prediction. For example, a fusion algorithm that accepts highly processed data containing each sensor's best target discrimination estimate can be the optimal one to use for the detection and classification problem when each sensor responds to different signature-generation phenomena. But another fusion algorithm that accepts minimally processed data from more than one sensor and then analyzes and associates these data to form tracks can be shown to be optimal for obtaining the most accurate track position estimates.

3.2.1 Detection, classification, and identification algorithms for data fusion

A taxonomy for detection, classification, and identification algorithms used in Level 1 processing is shown[2,3,9,10,11] in Figure 3.3. The major algorithm categories are physical models, feature-based inference techniques, and cognitive-based models. Mathematical approaches for data fusion, other than those listed in the figure, have been developed in recent years. These include random set theory, conditional algebra, and relational event algebra.[12] Random set theory deals with random variables that are sets rather than points. Goodman et al.[12] have used random set theory to reformulate multisensor, multitarget estimation problems into single-sensor, single-target problems. They have also applied the theory to incorporate ambiguous evidence (e.g., natural language reports and rules) into multisensor, multitarget estimation and to incorporate various expert system methods (e.g., fuzzy logic and rule-based inference) into multisensor, multitarget estimation. Conditional event algebra is a type of probabilistic calculus suited for contingency problems such as knowledge-based rules and contingent decision making. Relational event algebra is a generalization of conditional event algebra that provides a systematic basis for problems involving pooling of evidence.

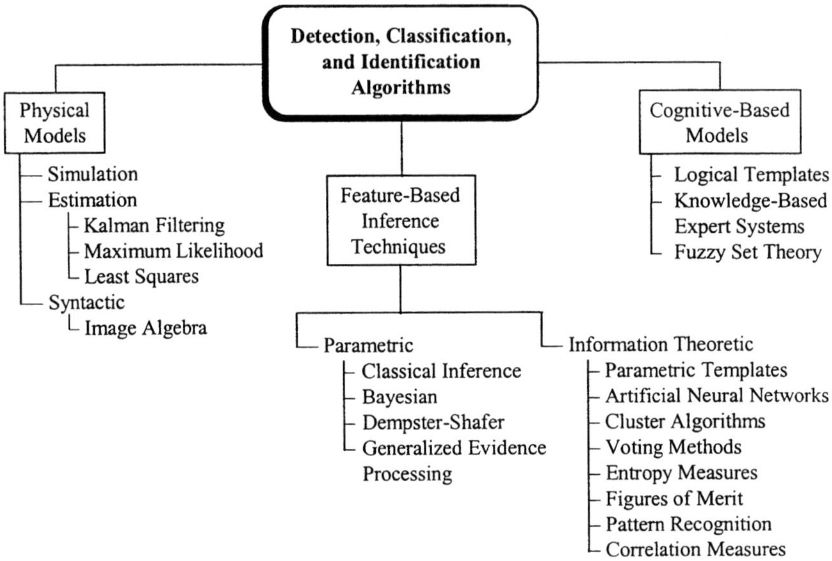

Figure 3.3: Taxonomy of detection, classification, and identification algorithms
(Adapted from Refs. 2, 3, 9-11)

Physical models — Physical models replicate object discriminators that are easily and accurately observable or calculable. Examples of discriminators are radar cross section as a function of aspect angle; IR emissions as a function of vehicle type, engine temperature, or surface characteristics such as roughness, emissivity, and temperature; and height profile images. Physical models estimate the classification and identity of an object by matching modeled or prestored target signatures to the observed data. The techniques listed in this category include simulation, estimation, and syntactic methods. The estimation processes, such as Kalman filtering, maximum likelihood, and least squares approximation, are analogous to the state estimation and tracking algorithms described in Section 3.2.2. The syntactic methods, although shown here, are described later as part of pattern recognition listed under information theoretic techniques.

Feature-based inference techniques — Feature-based inference techniques perform classification or identification by mapping data, such as statistical knowledge about an object or recognition of object features, into a declaration of identity. Feature-based algorithms may be further divided into parametric and what Waltz and Llinas refer to as information theoretic techniques as depicted in Figure 3.3.

Parametric techniques

Parametric classification directly maps parametric data (e.g., features) into a declaration of identity. Physical models are not used. Parametric techniques include classical inference, Bayesian, Dempster-Shafer, and generalized evidence processing.

Classical inference gives the probability that an observation can be attributed to the presence of an object or event, given an assumed hypothesis. Its major disadvantages are: (1) difficulty in obtaining the density function that describes the observable used to classify the object, (2) complexities that arise when multivariate data are encountered, (3) its capability to assess only two hypotheses at a time, and (4) its inability to take direct advantage of *a priori* likelihood probabilities.

Bayesian inference resolves some of the difficulties with classical inference. It updates the likelihood of a hypothesis given a previous likelihood estimate and additional observations and is applicable when more than two hypotheses are to be assessed.[11,13] The disadvantages of Bayesian inference include: (1) difficulty in defining prior likelihood functions, (2) complexities when there are multiple potential hypotheses and multiple conditionally dependent events, (3) mutual exclusivity required of competing hypotheses, and (4) inability to account for general uncertainty. Bayesian inference is discussed further in Chapter 4.

The Dempster-Shafer method generalizes Bayesian theory to allow for uncertainty by distributing support for a proposition (e.g., that an object is of a particular type) not only to the proposition itself, but also to the union of propositions (disjunctions) that include it. Any support that cannot be directly assigned to a proposition, or its negation, is assigned to the set of all propositions making up the hypothesis space (i.e., uncertainty). Bayesian and Dempster-Shafer produce identical results when all the singleton propositions are mutually exclusive and there is no support assigned to uncertainty. Disadvantages of the Dempster-Shafer method include the need to define the processes in each sensor that assign the degree of support for a proposition. Examples of the Dempster-Shafer approach to data fusion are given in Chapter 5.

Generalized evidence processing (GEP), which separates the hypotheses (propositions) from the decisions, allows Bayesian decisions to be extended into a hypothesis space (called the frame of discernment in Dempster-Shafer theory) that incorporates multiple hypotheses (as with Dempster-Shafer). With GEP, evidence from nonmutually exclusive propositions can be combined in a Bayesian formulation to reach a decision.[14,15] As with Dempster-Shafer, GEP uses evidence collected by multiple sensors and assigns probability mass to the evidence. GEP diverges from Dempster-Shafer in that the probability mass assignments and their combination are based on the *a priori* conditional probability of the propositions or hypotheses.[11]

Information theoretic techniques

Information theoretic techniques transform or map parametric data into an identity declaration. All these methods share a similar concept, namely, that similarity in identity is reflected in similarity in observable parameters. No attempt is made to directly model the stochastic aspects of the observables. The techniques that can be included under this category are parametric templates, artificial neural networks, cluster algorithms, voting methods, entropy-measuring techniques, figures of merit, pattern recognition, and correlation measures.

In *parametric templating*, multisensor data acquired over a period of time and multisource information are matched with preselected conditions to determine if the observations contain evidence to identify an entity. Templating can be applied to event detection, situation assessment, and single object identification.[3]

Artificial neural networks are hardware or software systems that are trained to map input data into selected output categories. The transformation of the input data into output classifications is performed by artificial neurons that attempt to emulate processes that occur in biological nervous systems. Artificial neural networks are discussed in detail in Chapter 6.

Cluster algorithms group data into natural sets or clusters that are interpreted by an analyst to see if they represent a meaningful object category. All cluster algorithms require a similarity metric or association measure that describes the closeness between any two feature vectors, for example, one that represents the input data and one that represents a potential class to which the data belong. Cluster algorithms operate with five basic steps: (1) selection of sample data, (2) definition of the set of variables or features that characterize the entities in the sample, (3) computation of the similarities among the data, (4) use of a cluster analysis method to create groups of similar entities based on data similarities, and (5) validation of the resulting cluster solution. The application of cluster algorithms may lead to biased results because of the heuristic nature of these algorithms. In general, data scaling, choice of similarity metric and algorithm, and sometimes even the order of the input data may substantially affect the resulting clusters. Hence, application of cluster methods must be judged on their effectiveness and ability to form consistent and meaningful identity clusters.[3]

Voting methods combine detection and classification declarations from multiple sensors by treating each sensor's declaration as a vote in which majority, plurality, or decision-tree rules are used. Additional discrimination can be introduced via weighting of the sensor's declaration as discussed in Chapter 7 where voting based on Boolean algebra is described.

Entropy measures take their name from communications theory and attempt to measure the importance of the information in a message by its probability of occurrence. Frequently occurring messages or data are of low value, while surprising or rare messages are of higher value. The function that measures the value of information, therefore, has the property that it decreases with increasing probability of receiving the information.

Figures of merit are metrics derived from plausible or heuristic arguments that aid in establishing a degree of association between observations and object identity. They contain flexible sets of algorithms that measure the strength of entity and event relationships. Figure of merit techniques attempt to formulate a relationship among several variables in order to improve the association or classification of input data. Sometimes figures of merit are considered a templating approach since they reflect the expected observations, behaviors, logical relationships, and any other basis that profiles an object's identity.

Pattern recognition concerns the description or classification of data. Historically, the two major approaches to pattern recognition have been the statistical (or decision theoretic) and the syntactic (or structural). More recently, artificial neural networks have provided a third approach. In statistical pattern recognition, a set of characteristic measurements or features are extracted from the input data and used to assign the feature vector to one of *c* classes. Assuming features are generated by a state of nature, the underlying statistical model represents a state of nature, set of probabilities, or probability density functions that correspond to a particular class.[16] Syntactic pattern recognition is applied when the significant information in a pattern is not merely the presence or absence of numerical values, but rather the interconnections of features that yield structural information. The structural similarity of patterns is assessed by quantifying and extracting structural information using, for example, the syntax of a formally defined language. Typically, syntactic approaches formulate hierarchical descriptions of complex patterns from simpler subpatterns or primitives. Neural computing, as discussed above, attempts to mimic processes that occur in biological neural systems.

Pattern recognition is frequently applied to high-resolution, multipixel imagery such as that from a FLIR or high-resolution scanners found on satellites. Features extracted from a FLIR image may consist of temperature gradients, length/width ratios, central moments, and the relative size of subobjects within the boundary of the larger object. Features associated with LANDSAT images are extracted from each pixel of data for each spectral band in the sensor. Frequency domain spectra of MMW signatures also provide features used in statistical pattern recognition algorithms. Features in this case are extracted from the Fourier transformed signal. Schalkoff[16] provides a concise comparison of the attributes of the statistical, syntactic, and neural pattern recognition approaches as shown in Table 3.2.

Table 3.2: Comparison of statistical, syntactic, and neural pattern recognition (PR) approaches

Attribute	Statistical PR	Syntactic PR	Neural PR
Pattern generation (storing) basis	Probabilistic models	Formal grammars	Stable state or weight array
Pattern classification basis	Estimation/decision theory	Parsing	Based on properties of the neural network
Feature organization	Feature vector	Primitives and observed relations	Neural input or stored states
Typical learning (training) approaches			
Supervised:	Density or distribution estimation	Forming grammars (heuristic or grammatical inference)	Determining neural network system parameters (e.g., weights)
Unsupervised:	Clustering	Clustering	Clustering
Limitations	Difficulty in expressing structural information	Difficulty in learning structural rules	Often little semantic information from network

(R. Schalkoff, *Pattern Recognition: Statistical, Structural, and Neural Approaches*, John Wiley, NY [1992])

Correlation measures are derived from weighted combinations of figures of merit. They allow a comparison score or measure of correlation to be calculated for systems that have numerous figures of merit. Thus, the correlation measure represents the total likelihood that two entities are the same.

Cognitive-based models — Cognitive-based models, including logical templates, knowledge-based systems, and fuzzy set theory, attempt to emulate and automate the decision-making processes used by human analysts.

Logical templates

Templating, as the name suggests, is a concept where a predetermined and stored pattern is matched against observed data to infer the identity of the object or to assess a situation. Parametric templates that compare real-time patterns with stored ones can be combined with logical templates derived, for example, from Boolean relationships.[3] Fuzzy logic may also be applied to the pattern matching technique to account for uncertainty in either the observed data or the logical relationships used to define a pattern.

Knowledge-based expert systems

Knowledge-based systems incorporate rules and other knowledge from known experts to automate the object identification process. They retain the expert knowledge for use at a time when the human inference source is no longer available. Computer-based expert systems frequently consist of four logical sections: (1) a knowledge base that contains facts, algorithms, and a representation of heuristic rules; (2) a global database that contains dynamic input data; (3) a control structure or inference engine; and (4) a human-machine interface. The inference engine processes the data by searching the knowledge base and applying the facts, algorithms, and rules to the input data. The output of the process is a set of suggested actions that is presented to the end user.

Fuzzy set theory

Fuzzy set theory opens the world of imprecise knowledge or indistinct boundary definition to mathematical treatment. It facilitates the mapping of system state variable data into control, classification, or other output data. Its use of fuzzy associative memory allows a proposition to have a membership value in a given class ranging from 0 (definitely not a member) to 1 (definitely a member). Fuzzy set theory is intuitively appealing in that it permits uncertainties in knowledge or identity boundaries to be applied to such diverse applications as identification of battlefield threats, target tracking, and control of industrial and automotive processes. Unlike neural networks that sum throughputs, fuzzy systems sum outputs. Chapter 8 contains a detailed discussion of fuzzy set theory and fuzzy logic and illustrative examples.

3.2.2 State estimation and tracking algorithms for data fusion

A taxonomy for state estimation and tracking algorithms used in Level 1 processing is shown[2,3,9,11] in Figure 3.4. The processes required to perform the tracking function are represented, at the top level, by algorithms that (1) determine the search direction and (2) associate and correlate data and tracks. Association and correlation are further separated into data alignment; data and object correlation; and position, kinematic, and attribute estimation. The majority of this section is concerned with data and object correlation techniques.

Search direction — Direction tracking systems can be sensor (data) driven or target (goal) driven. In sensor-driven systems, target reports (consisting of combinations of range, azimuth, elevation, and range-rate measurement data) initiate a search through the track file for tracks that can be associated with the reports. Target-driven systems use a primary sensor for tracking and use the target track to direct other sensors to acquire data or search databases for reports that can be associated with particular tracks.

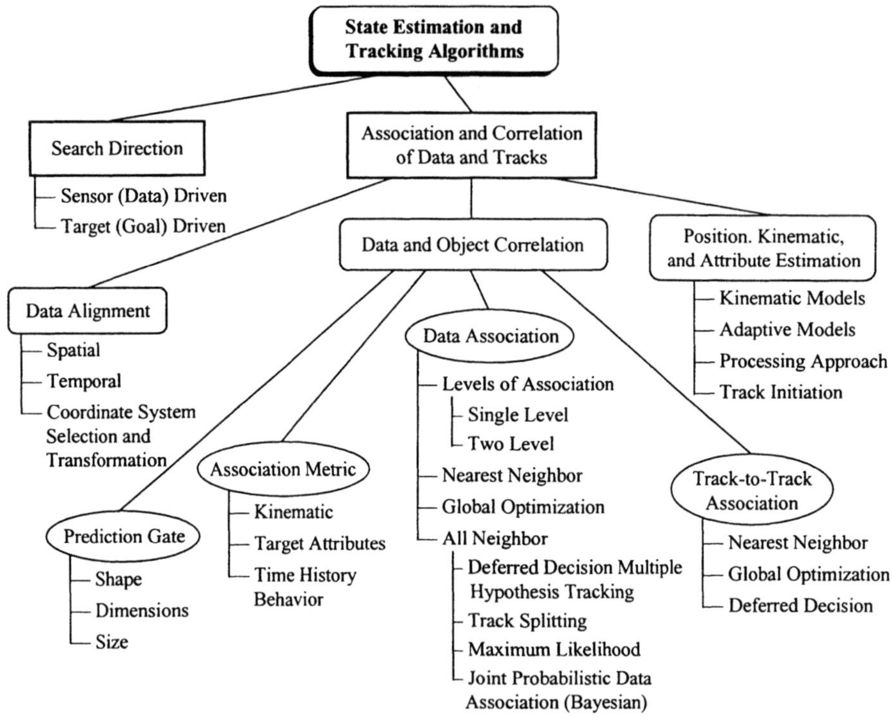

Figure 3.4: Taxonomy of state estimation and tracking algorithms
(Adapted from Refs. 2, 3, 9, and 11)

Association and correlation of measurement data and tracks — The proper association and correlation of measurement data and tracks from the multisensor inputs ultimately generates optimal central track files. Each file ideally represents a unique physical object or entity. Association and correlation require algorithms that define data alignment, prediction gates, association metrics, data and track association, and position, kinematic, and attribute estimation.

Data alignment

Data alignment is performed through spatial and temporal reference adjustments and coordinate system selection and transformations that establish a common space-time reference for fusion processing. Errors introduced by measurement accuracies, coordinate transformations, and unknown target motion are accounted for through this process.

Data and object correlation

Data and object correlation consist of processes that establish the prediction gate, define the association metric, perform data association, and perform track-to-track association.

Prediction gates

Prediction gates are established to control the association of data sets into one of two categories, namely, candidates for track update or initial observations for forming a new tentative track. Data that were originally categorized for track update may later be used to initiate new tracks if they are not ultimately assigned to an existing track. The size of the gates reflects the calculated or otherwise anticipated target position and velocity errors associated with their calculation, sensor measurement errors, and desired probability of correct association.

Data association metrics

Data association metrics quantify the similarity of the observations. Metrics are defined from the kinematic parameters (e.g., range, range rate, angle, position) and target attributes (e.g., temperature, size, shape, edge structure) that are observed and measured. The metric can be based on spatial distance (e.g., Euclidean distance) or statistical measures of correlation between observations and predictions (e.g., Mahalanobis distance), heuristic functions such as figures-of-merit that use the kinematic and target attribute information, and measures that quantify the realism of an observation or track based on prior assumptions such as track lengths, target densities, or track behavior.

Data association

In a multiple target and sensor scenario, data association refers to the statistical decision process that correlates reports from overlapping gates, multiple returns (hits) in a gate, clutter in a gate, and new targets that occur within a gate and successive scans. Association techniques that merge data and tracks from several sources into a single track usually employ either single-level tracking systems or two-level tracking systems. In a single-level tracking system, the data detected by the multiple sensors are transmitted to a single processing node (central-level fusion). Here the data are processed to initiate new tracks and update estimates of existing tracks. Association occurs between each new set of sensor data and the central track file.

Four variants of two-level tracking systems have been identified. The first (known as autonomous or distributed tracking) maintains separate sensor-level and central-level trackers. Each sensor-level tracker independently acquires, initiates, continues, and drops tracks based on its own data. In addition, track-to-track association is performed at a single node to form a central track file. The second two-level system uses either sensor data or sensor tracks to initiate and maintain a central track file. Track-to-track correlation is initiated at a central node by associating the sensor tracks to form a central track file. Sensors that provide information in the form of data rather than tracks send the data to the central node, where they are associated with the existing tracks or are used to establish new tracks. A third two-level system uses sensor data to form tracks in a central processor. The predicted gates for the next track update are sent from the central processor back to the sensors in order to cue their search area and velocity for the next update. A fourth two-level system associates sensor tracks in a central processor using track-to-track correlation to eliminate redundant tracks and to assign future reporting responsibility to the sensor with the best track (based on a covariance calculation or track quality assessment, for example). The data to be fused may originate from other command and control nodes or centers, as well as from sensors under common command and control.[17]

Data association can be accomplished with nearest-neighbor, global optimization, and all-neighbor techniques. In nearest-neighbor, a hard decision is made to pair the input data with the single best track. This process deteriorates when multiple, closely ranked choices and maneuvering targets are present. Several methods are available to remedy this shortcoming. One of these applies the Munkres or faster executing JVC (Jonker-Volgenant-Castanon) algorithm to globally optimize the association of all new data and tracks with any existing tracks.[18,19] An application of the JVC algorithm to the association of direction angle measurements is given in Chapter 9.

In all-neighbor association, several of the deficiencies of the nearest-neighbor approach are removed. In one implementation of all-neighbor association, called deferred decision multiple hypothesis tracking, each candidate pairing is considered a viable hypothesis and is retained in the track file until a decision criterion can eliminate or confirm the hypothesis. Final assignment of data is deferred until sufficient information from future scans is available to increase confidence in the hypothesis. When track association is deferred, however, the operator does not see the recommended track until several scans have elapsed. If the tracks are displayed for each scan, then the operator can potentially see multiple tracks, some of which are false, making situation assessment difficult. This deficiency has been overcome with new techniques that display only the high-confidence tracks.[20]

A variation of multiple hypothesis tracking, called track splitting, associates each report in the gate with a track, but does not specifically generate "new" tracks, nor compute the probability of correct association. The track splitting technique can be applied when a target maneuver is suspected. In this situation, the expected sensor update data may not be present in the normal gate. Therefore, the gate is enlarged to account for the maximum anticipated target maneuver. If the target is located within the larger gate, then the track is split into two parts, one corresponding to a non-maneuvered target and one to a maneuvered target. The decision to abandon one track or the other is made on the following scan.

Unlike other all-neighbor association techniques, maximum likelihood selects the most likely single target report for association with a track. Probability density functions are assumed for the target data, target track reports, and the spurious data reports due to noise, clutter, or decoys. A target is declared present if the likelihood function defined by the product of the probability density functions for the true and false targets is greater than a predetermined threshold.

Another form of all-neighbor association is a Bayesian approach called joint probabilistic data association (JPDA). JPDA is applicable to tracking multiple targets in scenarios with or without clutter. A related technique, probabilistic data association (PDA), is used for tracking single targets.[21] In these methods, each candidate pairing is used to update the track estimator, weighted by a quantitative factor that describes its probability of being correct.[22] All neighboring measurements contribute to the track; hence deferred decision making is not required. Bayesian techniques provide an optimal decision function when the *a priori* probabilities can be confidently established.

Track-to-track association

Track-to-track association is used to merge sensor-level tracks and thereby obtain a central track file. Tracks can be characterized by position, velocity, covariance, and other features. In order to associate the sensor-level tracks, they first have to be transformed into a common coordinate system and time aligned, as was discussed under data alignment. Gates are then formed and a metric is chosen to evaluate the track association process. Many of the methods discussed for data association can be used to perform track-to-track association. These include nearest neighbor, global optimization, and deferred decision. The latter operates on tracks obtained over several future scans. After the track associations are made, the target position and covariance matrix corresponding to the input tracks are combined to form a new target position and covariance for the fused track. If the states observed by the various sensors are not identical, then only those that are common are used in the association process. The remaining states are augmented to the track and carried along. Subsequent track association can be simplified by storing associated sensor track numbers. As updated tracks arrive from the sensors, the previous track associations are then simply verified before the global track file is updated.

The variation and complexity of the tracking problem, as categorized by single target–single sensor, single target–multiple sensor, multiple target–single sensor, and multiple target–multiple sensor, dictate the data and track association technique as suggested by Table 3.3. The method of association shown is generally appropriate for the given tracking complexity. Of course, the more complicated association techniques can be used for the single target cases as well. Furthermore, in cases where the sensor cannot adequately resolve targets within the gate, groups of targets may be tracked rather than individual targets.

Table 3.3: Suggested data and track association techniques for different levels of tracking complexity

Tracking Complexity	Association Technique	Number of Scans
Single target–Single sensor	Nearest neighbor	Single
	Multiple hypothesis tracking	Multiple
	Track splitting	Multiple
Single target–Multiple sensor	Nearest neighbor	Single
	Multiple hypothesis tracking	Multiple
	Track splitting	Multiple
Multiple target–Single sensor	Nearest neighbor	Single
	JVC	Single
	Multiple hypothesis tracking	Multiple
	Track splitting	Multiple
	Maximum likelihood	Single or Multiple
	JPDA	Single
Multiple target–Multiple sensor	Nearest neighbor	Single
	JVC	Single
	Multiple hypothesis tracking	Multiple
	Track splitting	Multiple
	Maximum likelihood	Single or Multiple
	JPDA	Single

Positional, kinematic, and attribute estimation

These processes optimally combine multiple observations to obtain improved estimates of the position, velocity, and attributes (e.g., size, temperature, and shape) of an object. Estimates of updated target parameters are provided by a tracking filter. The filter uses algorithms based on time sequences of associated measurements to develop predictions of target state and attributes. Kinematic and adaptive models of object motion and sequential or batch processing (i.e., where all data are processed simultaneously) techniques are used to support the estimation process. The estimators also include *a priori* models of track dynamics and observations to refine the state estimate and to predict the state at the next observation interval for gating.

Even with *a priori* knowledge, the target may maneuver. Therefore, the state of the tracking filter must be changed to accommodate the maneuver. This can be accomplished in several ways. The first, used with track splitting, augments the state of the parent track to include the maneuver. The second, called multiple model maneuver, parameterizes the range of the expected maneuver and constructs tracking filters for each set of parameter values. Bloom and Bar-Shalom assume a transition probability for each of the sets of parametric values used to construct the filters.[23] States that are incorporated into filters must correspond to the observables of the tracking sensor. For example, if the state of a tracker is selected as position, velocity, and attitude (pitch, roll, and yaw), but only azimuth, elevation, and range are measured, then the attitude is not observable and the state cannot be updated.

Track initiation

Several methods of track initiation are available to acquire targets and begin the tracking process. The simplest method uses single scan association to establish a detection gate based on minimum and maximum anticipated target speeds. When a detection not associated with another track is made, a gate is centered about the detection coordinates. Detections made on subsequent scans within the gate are then associated with the first detection. A track is initiated for every possible pairing of the first detection with subsequent ones. Usually detections on two consecutive scans are required to initialize the Kalman state and covariance track estimation filter for position and velocity. By limiting the association of detections to those on two consecutive scans, the gate size is minimized for the second detection and, thus, the creation of false tracks is minimized.

The promotion of the initiated tracks to system tracks is based on rules such as "n out of m." Here n detections out of m scans are required to declare the track a system track. Values of n and m are established from requirements that specify the number of false tracks, probability of target detection, clutter density, and the time allowed to declare a track. Another method of track initiation applies the maximum likelihood algorithm to several scans of stored data to maximize the probability of correctly associating the detections. In this case, processor capabilities usually limit the number of scans that are compared.[24]

3.3 Level 2, 3, and 4 processing

Once the Level 1 or low-level processing is complete, the target identities and tracks are input to the higher level situation assessment (Level 2) and threat assessment (Level 3) fusion processes. Refinement of the fusion process itself (Level 4) occurs through process evaluation and control that includes guidance for the acquisition of new data.

According to the Data Fusion Development Strategy Panel, Level 2 processing identifies the probable situation causing the observed data and events. Thus, it develops a description or interpretation of the current relationships among fixed and moving objects and events in the context of the operational environment. The data obtained from Level 1 analysis are now used to gain insights into prescribed event and activity sequences, force structures, and the overall battle environmental factors. Key functions of Level 2 processing, in terms of a military application, include:

- Object aggregation – establishing relationships among objects including temporal, geometrical proximity, communications links, and functional dependence;

- Event and activity aggregation – establishing temporal relationships among diverse entities to identify meaningful events or activities;

- Contextual interpretation and fusion – analyzing data with respect to the context of the evolving situation including weather, terrain, sea state or underwater conditions, enemy doctrine, and socio-political considerations.

Level 3 processing develops a threat-oriented data perspective to estimate enemy capabilities, identify threat opportunities, estimate enemy intent, and determine levels of danger. Threat assessment is distinct from situation assessment in that threat assessment includes multiperspective and quantitative enemy force analyses needed to estimate the enemy's course of action and force lethality. The critical functions that support threat assessment include:

- Capability estimation – predicting the size, location, and capabilities of enemy forces;

- Prediction of enemy intent – determining enemy intention based on actions, communications, doctrine, culture, history, education, and political structure;

- Identification of threats – identifying potential threat opportunities based on prediction of enemy actions, operational readiness analysis of friendly vulnerabilities, and analysis of environmental conditions;

- Multiperspective assessment – analyzing the data with respect to the friendly, enemy, and neutral perspectives, including effects of time and space on force deployment and preparing estimates of the enemy war plan;

- Offensive and defensive analysis – predicting the results of hypothesized enemy engagements considering rules of engagement, enemy doctrine, and weapon models.

Level 4 processing monitors and evaluates the ongoing fusion process to refine the process itself and to regulate the acquisition of data to achieve optimum results. Fusion process refinement interacts with each of the other levels and with external systems or the system operator. Its key functions include:

- Evaluations – assessing performance and effectiveness of the fusion process to establish real-time control and long-term process improvements;

- Fusion control – identifying changes or adjustments to processing functions within the data fusion domain that may result in improved performance;

- Source requirements processing – determining source-specific data requirements (specific sensors, sensor data, qualified data, reference data, etc.) needed to improve the multilevel fusion products;

- Mission management – recommending allocation and direction of resources (sensors, platforms, communications, etc.) to achieve overall mission goals.

Large databases, including the ability to support fast data insertion and fast data retrieval, are often needed to develop the higher level fusion processes. The databases are maintained by management systems that provide monitoring, evaluation, addition, updating, retrieval, merging, and purging of data. Time tagging of entries is a technique used to aid the relevance of inferences drawn from these databases.

Figure 3.5 illustrates a command and control architecture as might be used in a military application to combine sensor data with information from a variety of diverse sources. The operational environment represented by the circle on the left side of the figure contains data entries that aid target identification and tracking, as well as situation and threat assessment found in Level 1, 2, and 3 fusion. The information that typically supports these fusion processes is detection and tracking data from land, air, sea, and space-based sensors including friendly missile guidance data from Global Positioning System satellites; lethality estimates; force/weapon composition; targeting ability; order of battle; and alert status for enemy and friendly forces. Additional information is provided by weather sensors, diplomatic messages, analysis of political and economic factors, and other intelligence. The middle of the figure depicts Level 1 data fusion of real-time sensor data and historical database entries in support of target identification and tracking. Data from similar and dissimilar sources have been isolated to indicate that unique processing may be required for each type of information. Additional databases supply information to the Level 2 and 3 situation and threat assessment processes shown on the right. A database management system (DBMS) supports database housekeeping functions. The nodal interconnectivity boxes indicate that processing may occur both within a processing node and across processing nodes. Finally, the term "dynamic, integrated situation representation" represents the changeable nature of military

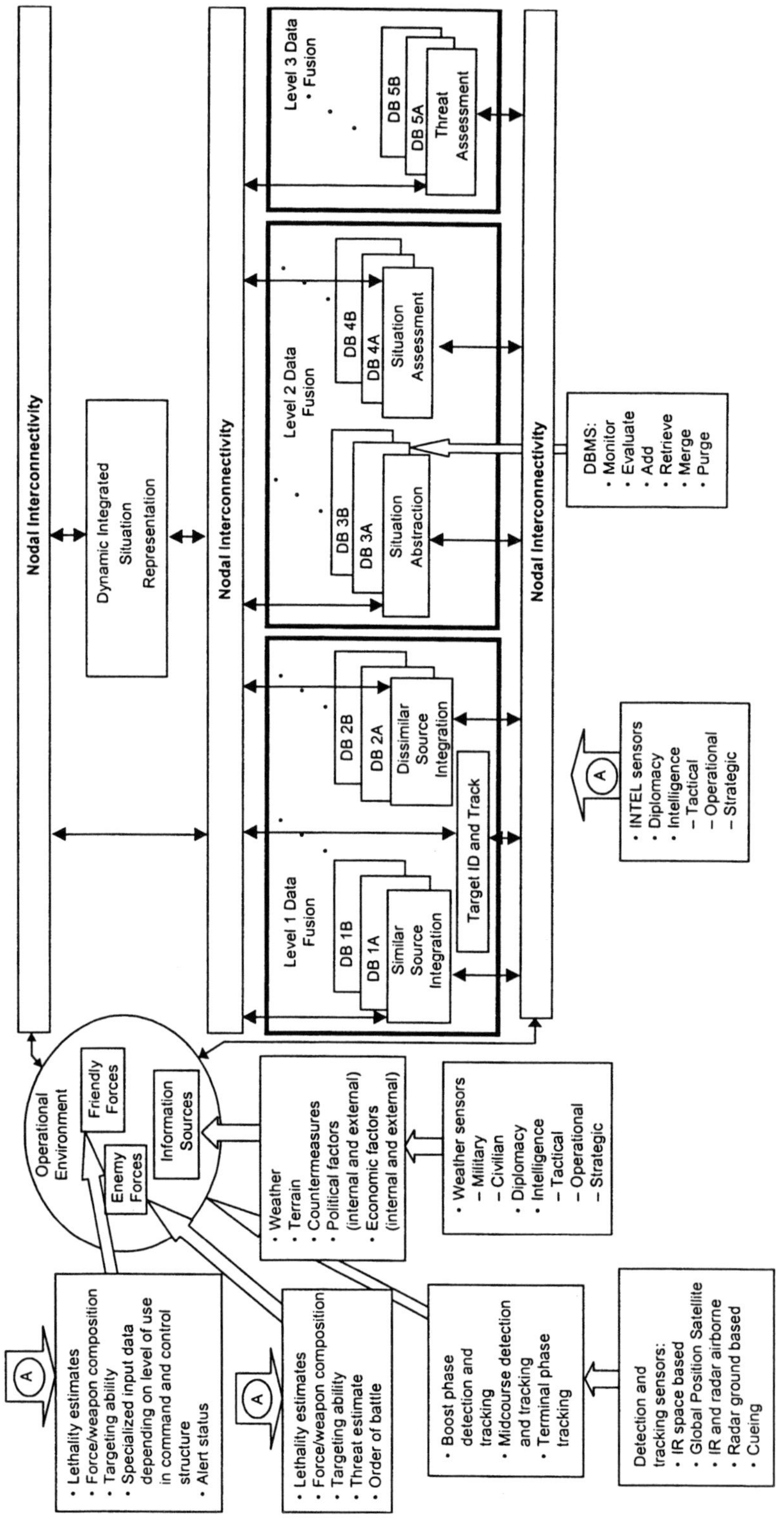

Figure 3.5: Military command and control system architecture showing fusion of information from multiple sources at multiple locations

environments and the dependence of the fusion results on the synthesis of information from diverse and multilevel sources.

3.4 Definition of an architecture

An architecture is a system of components whose structure and integration enable it to perform functions that the individual components could not otherwise accomplish. An architecture initially provides conceptual design information to develop cost and operational effectiveness analyses, risk analyses, and technology transition. Design information includes specification of the components and their interconnections, data and information flows, system operating modes, and allocation of functions and subfunctions to particular architecture components and to alternates that assume the functions of failed components. The architecture identifies production, test, and support requirements and determines design constraints for configuration items (i.e., a system element or an aggregation of system elements that performs an end-use function and is designated for configuration control). As the architecture matures, it provides preliminary and detailed design information for system elements and their integration into products and processes.[25,26] As shown in the sections below, the definition of a data fusion architecture fits within the framework laid out in the broader architecture definition.

3.5 Data fusion processes

Before examining various types of data fusion architectures, it is worthwhile to formally define several processes that occur in the fusion processor. Many of these were introduced in earlier sections. The fusion processor analyzes the inputs from all the sensors and performs the alignment, association, correlation, estimation, classification, and cueing functions defined below:[27]

- Alignment – referencing of sensor data to a common time and spatial origin;

- Association – using a metric to compare tracks and measurement data (reports) from different sensors to determine candidates for the correlation process;

- Correlation – combining tracks and measurement data that are matched during association to improve detection, classification, and tracking of objects of interest;

- Estimation – predicting an object's future position by updating the state vector and error covariance matrix using the results of the correlation process;

- Classification – assessing the tracks and object discrimination data to determine target type, lethality, and threat priority;

- Cueing – feedback of threshold, integration time, and other data processing parameters or information about areas over which to conduct a more detailed search, based on the results of the fusion process. For example, if a region of high clutter is found, a signal may be sent to the appropriate sensor to increase the threshold setting. Alternatively, when the fusion processing identifies a decoy, a message describing the decoy's location is sent to minimize target search-related signal processing in this region. Another application of cueing is to initiate a search of a small, but high-interest region using a sensor of limited field of regard having high resolution, such as a laser radar.[28]

3.6 Data fusion architectures

There are several ways to classify data fusion architectures. In one approach, the architecture is defined by the extent of the data processing, the data products produced by the individual sensors, and the location of the fusion processes. For example, sensors supplying information to detection, classification, and identification fusion algorithms may utilize complex processing techniques to provide the object class to a fusion algorithm for further refinement. Alternatively, the sensors may simply provide filtered signals or features to a fusion algorithm, where the signals or features are analyzed in conjunction with those from other sensors to determine the object class. On the other hand, sensors supplying information to state estimation and tracking algorithms may provide either measurement data, i.e., reports, that contain the position and velocity of objects or tracks of the objects. Current values of measurement data may be combined with previously obtained data to generate new tracks or the current data may be used to update pre-existing tracks using Kalman filtering. These processes can occur in the individual sensors or at a central processing node, depending on the architecture. If the sensors supply tracks, the tracks can be correlated with pre-existing tracks residing in individual sensors or at a central processing node.

The terms that describe architectures based on the extent of the data processing, data product types, and fusion location are sensor-level fusion (also referred to as autonomous fusion, distributed fusion, and post-individual sensor processing fusion), central-level fusion (also referred to as centralized fusion and pre-individual sensor processing fusion), and hybrid fusion, which uses combinations of the sensor-level and central-level approaches.[29,30,31] The resolution of the data and the extent of the processing may also be employed to define another fusion lexicon. The nomenclature used in this case is pixel-level, feature-level, and decision-level fusion.

3.6.1 Sensor-level fusion

The sensor-level fusion architecture is illustrated in Figure 3.6. It may be shown to be optimal for detecting and classifying objects if the sensors use independent signature-generation phenomena to develop information about the identity of objects in the field of regard, i.e., they derive object signatures from different physical processes and generally do not false alarm on the same artifacts.[32] The sensor footprints must also be registered with respect to each other to ensure that the sensors are describing the characteristics of events or objects at the same spatial locations.

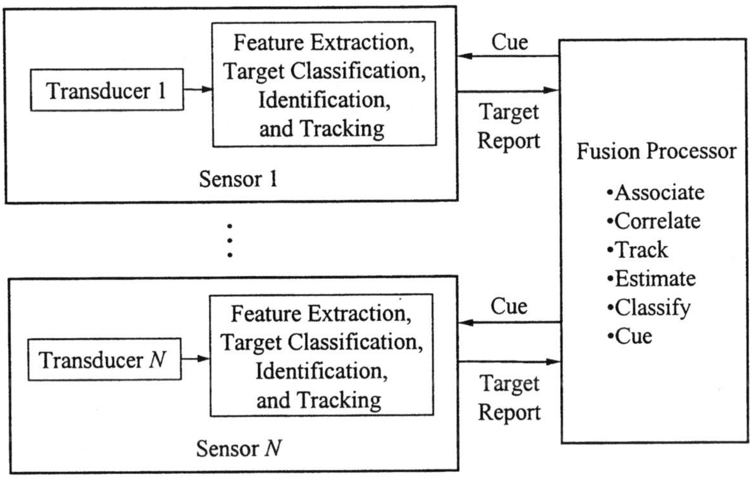

Figure 3.6: Sensor-level fusion

Phenomena that generate the signatures detected by various types of sensors are shown in Table 3.4. Acoustic sensor signatures have been included since they are frequently used in military and transportation applications.

The signatures are not only a function of the objects and background, but also of the sensor design parameters as illustrated in Table 3.5. The signatures received by active sensors are subject to the transmitted frequency and polarization, waveform shape, and power, while signatures from passive sensors are not. The target shape, size, material, small-scale structure, orientation, and relative motion are other factors that influence the signature detected by active sensors. The emissivity, roughness, incidence angle, temperature of the surface, and receiver polarization affect the passive signature. Background properties such as clutter type, weather and other atmospheric effects, and the presence of countermeasures affect the signatures presented to active and passive sensors by absorbing and scattering energy associated with real targets and by creating false target signatures.

Table 3.4: Signature-generation phenomena

Sensor	Detectable Signature	Signature Dependencies
MMW radar	Radar cross section	Shape, material composition, surface smoothness and regularity, gaps, cavities, receiver polarization
MMW radiometer	Apparent temperature	Emissivity and temperature of object, receiver polarization and incidence angle, surface roughness, weather, atmospheric conditions
Infrared FLIR or IRST	Emission and reflectance	Radiance produced from within the object, such as by engines, and radiance produced by natural sources, such as direct heating by the sun or by reflected radiation
Visible	Reflection	Weather, atmospheric conditions, contrast with the background, visible emissions from exhausts
Electronic support measures (ESM)	Electronic emissions	Active sensor and transmitter sources such as communications equipment, navigation and guidance systems, fire control systems, electronic countermeasures, and, in general, any other source of electromagnetic radiation
Magnetic	Perturbation in Earth's magnetic field or change in an induced magnetic field	Magnetism associated with ferromagnetic materials (dipoles aligned parallel to their neighbors) and ferrimagnetic materials or ferrites (neighboring dipoles are aligned in an antiparallel arrangement, but different types of dipoles are present and the dipoles do not cancel)[33]
Acoustic	Acoustic energy	Engine noise, noise of object as it moves through air or moves on the ground surface such as produced by an airframe or ground vehicle

Several types of signature-generation phenomena can be exploited in a multiple sensor system. A passive IR sensor develops signatures from differences between the absolute temperatures and emissivities of the objects and background in the field of view. The emissivities are dependent on the surface characteristics of the particular object and the wavelength band in which the sensor operates. But laser radar can function as a multiple phenomena-sensing device in its own right. It receives a portion of the transmitted energy scattered from the objects and background that is proportional to their reflectance and scatterer shape and size. It also receives range data from which the distance to the scatterers can be calculated.

Microwave and MMW radars receive a portion of the transmitted energy scattered from the objects and background that is proportional to the size of the scattering cross section of the object. Radars with larger fields of regard are capable of scanning the required search area faster than the infrared wavelength sensors, but with less resolution. However, the microwave and MMW radars operate in rain, fog, haze, clouds, and smoke with less absorption than the IR.

Table 3.5: Sensor, target, and background parameters that contribute to object signature characterization

Sensor	Target	Background
• Spatial resolution	• Shape	• Clutter distribution
• Transmit and receive frequencies	• Overall physical size	• Clutter magnitude
• Frequency stability	• Small-scale structure	• Clutter decorrelation time
• Polarization	• Gross and small-scale signature parameters	• False targets and sun glint
• Transmit waveform	• Orientation	• Jammers
• Transmit power	• Number and relative positions	• Rain
• Scanning mechanism		• Smoke
• Noise figure		• Dust
• Sensitivity		• Haze
• Bandwidth		• Fog
• Operating range		• Clouds
• Data registration		

Once the sensor system designer is assured that the sensor selection will provide signatures based on independent phenomena, the sensor outputs can be combined in a sensor-level fusion architecture. The outputs from the sensors are fed into a fusion processor after the input data have been optimally processed by each sensor. The signal processing can thus be optimized for each sensor according to its resolution, frequency, field of regard, scan rate, and other attributes. Time-domain processing can be used for one sensor, frequency-domain techniques with another, and multipixel image processing algorithms with a third.

In detection, classification, and identification fusion, two pieces of information must be present in each sensor's output to the fusion processor. These are (1) the detection, classification, or identification decision and (2) how well, or with what confidence, the sensor has been able to detect, classify, or identify the objects in the field of regard. When tracking is of interest, a third piece of information is required, namely the location of the object or its track. With these inputs, it is possible to design a fusion algorithm that can combine the sensor data and improve upon the decision made by any sensor acting alone. In fact, sensor-level fusion can be shown to be as optimal (based on Bayesian decision logic) for detecting, classifying, and identifying targets as central-level fusion, which relies on minimally processed sensor data, when the sensors derive their information from independent signature-generation processes.[32] Three sensor-level fusion approaches, Bayesian inference, Dempster-Shafer inference, and the voting fusion based on Boolean algebra, are discussed in detail in later chapters.

3.6.2 Central-level fusion

Figure 3.7 depicts the central-level fusion architecture. In detection, classification, and identification data fusion, each sensor may provide minimally processed data to the fusion processor. Minimal processing includes operations such as filtering and baseline estimation. In state estimation and tracking fusion, the sensors typically supply measurement data, although sensor-generated tracks may also be sent to the fusion processor.

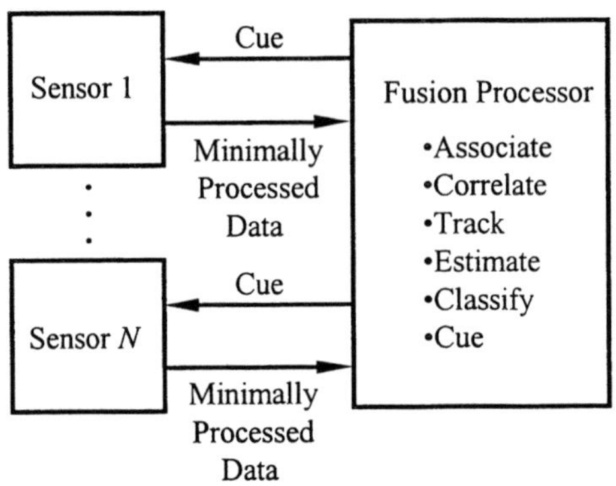

Figure 3.7: Central-level fusion

Central-level fusion algorithms are generally more complex and must process data at higher rates than in sensor-level fusion, because the centralized architecture is designed to operate on the minimally analyzed data output by each sensor. The central-level fusion algorithm examines input data for target features or attributes that aid in tracking and discriminating among objects. Central-level fusion is optimal for tracking objects, performing more effectively than sensor-level fusion in estimating or predicting the future position of the object. Blackman observes that the increased tracking accuracy is due to a combination of effects: (1) processing all the data in one place, (2) forming the initial tracks based on observations from more than one sensor, thus eliminating tracks established from partial data received by the individual sensors, (3) processing sensor reports directly, eliminating difficulties associated with combining the sensor-level tracks produced by the individual sensors, and (4) facilitating multiple hypothesis tracking by having all data available in a central processor.[31] Deficiencies of the method are reflected in the large amount of data that must be transferred in a timely manner to the central processor(s) and then be processed by them. Central-level fusion target tracking and discrimination algorithms can be written to tolerate lack of particular sensor inputs. The advantages of sensor-level and central-level fusion are compared in Table 3.6. The hybrid fusion algorithm discussed next can be used to combine both target tracks and measurement data from multiple sensors.

Table 3.6: Comparative advantages of sensor-level and central-level fusion

Sensor-Level Fusion	Central-Level Fusion
• Discrimination among potential targets or objects of interest before data entry into the fusion processor reduces the load on the fusion processor	• More accurate object discrimination than with sensor-level fusion, if the multisensor data are not generated by independent phenomena
• Optimization of each sensor's signal processing to the nuances of the transducer design and kinematics	• Optimization of object track and position estimates
• Cueing to adjust sensor signal processing or search area parameters based on data from other sensors	• Reduced weight, volume, power, and production cost in comparison with sensor-level fusion, if fewer processors are used
• Flexibility in the numbers and types of sensors to allow addition, removal, or substitution of sensors without having to alter the fundamental structure of the fusion algorithm	• Increased reliability of signal processing hardware, if fewer processors are used overall to support the fusion algorithms; reliability can be increased further, if required, by providing redundant paths for the processing
• Cost-effective alternative for adding data fusion into an existing multi-sensor configuration	

3.6.3 Hybrid fusion

In hybrid fusion, shown in Figure 3.8, the central-level fusion process is supplemented by individual sensor signal processing algorithms that may, in turn, provide inputs to a sensor-level fusion algorithm as well. Hybrid fusion allows the tracking benefits of central-level fusion to be realized using the sensor measurement data and, in addition, allows fusing of target tracks as computed by the individual sensors in the sensor-level fusion process.

Global track formation that combines the central- and sensor-level fusion tracks occurs in the central-level processor. Hybrid fusion can also be used to support target attribute classification when the signature data are not truly generated by independent phenomena. In this case, minimally processed data are sent to a central processor where they are combined using a fusion algorithm to detect and to classify the objects in the field of view of the sensors. The disadvantages of hybrid fusion are the increased processing complexity and possibly increased data transmission rates.

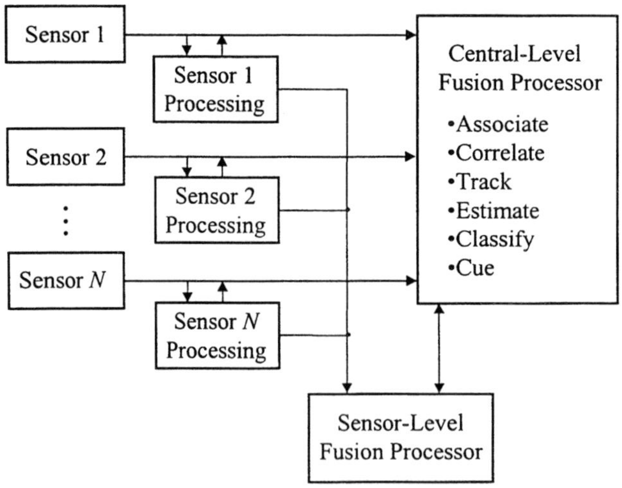

Figure 3.8: Hybrid fusion

3.6.4 Pixel-level fusion

In pixel-level fusion, minimally processed data are combined at the pixel or resolution-cell level of the sensors using a central-level fusion architecture. Data from different sensors, or sensor channels within a common sensor, are combined before much preprocessing is performed.

Pixel-level fusion is applied to LANDSAT imagery to detect diseased crops or identify a particular crop. Identification is not made using the individual spectral bands of data, but rather the information from all bands is combined in a pixel-level fusion process before the scene is identified.

Figure 3.9 illustrates another example of pixel-level fusion using CO_2 laser radar data. Range data from objects in the background and foreground (upper left image) are combined with intensity data characteristic of the reflectance of the objects in the background and foreground (upper right image). Although the nature of the background object is not clear to an observer viewing the separate range or intensity images, the object becomes apparent once the data are fused at the pixel level. At this stage, human or automatic target classification algorithms may be applied.[32,34] It is possible to encounter image or data registration problems when fusing data from different sensors. In this laser radar, however, the pixels in the range and intensity images are perfectly aligned because they are produced by the same sensor.

Range Image

Intensity Image

Fused Image

Figure 3.9: Pixel-level fusion in a laser radar

(G.S. Robinson and A.O. Aboutalib, "Trade-off analysis of multisensor fusion levels," *Proc. of the 2nd Nat'l. Sym. on Sensors and Sensor Fusion*, Vol. II, GACIAC PR 89-01, 21-34, IIT Res. Inst., Chicago, IL, [1990])

3.6.5 Feature-level fusion

Feature-level fusion is used either in central-level or sensor-level fusion architectures. Features are extracted from each sensor and combined into a composite feature, representative of the object in each sensor's field of view. An example of a composite feature is one constructed by stringing individual sensor feature vectors end-to-end (concatenation) to form a longer vector that serves as the input to a classifier.

Another example of feature-level fusion occurs with multilayer artificial neural networks as depicted[35] in Figure 3.10. As shown here, target features are extracted from a millimeter-wave radar, passive infrared sensor, and laser radar. The features are combined to form a composite vector that is input to a neural network. The network, trained off-line to recognize the targets of interest and differentiate them from false targets, assigns observed objects to particular classes with some probability, confidence, or priority. Training is performed using all sensor data simultaneously. Therefore, if one of the sensors is replaced by a different sensor type, simultaneous data collection and training have to be repeated.

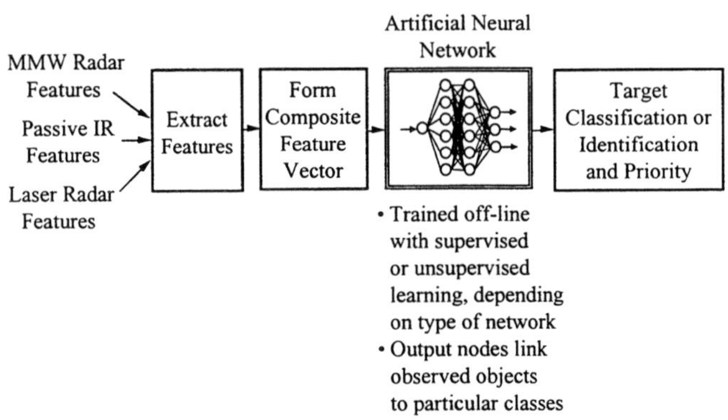

Figure 3.10: Feature-level fusion in an artificial neural network classifier

3.6.6 Decision-level fusion

Decision-level fusion is associated with sensor-level fusion as shown in Figure 3.6. The results of the initial object detection and classification by the individual sensors are input to a fusion algorithm. Final classification occurs in the fusion processor using an algorithm that combines the detection, classification, and position attributes of the objects located by each sensor. Classification performance is suboptimal compared to that of feature-level fusion unless the sensors respond to independent signature-generation phenomena.[32]

3.7 Sensor footprint registration and size considerations

When sensors are located at different spatial positions or, for that matter, collocated on the same platform, it is desirable to have their footprints overlap in target-detection space. Furthermore, the data from each sensor must be time and spatially aligned, or registered, with respect to those from the other sensors. Overlapping sensor footprints ensure that time-dependent phenomena (such as clutter decorrelation or target motion) are observed by all sensors at the same time. This footprint configuration thus supports optimal fusion of the sensor data within the overlapping fields of view. If data from overlapping sensors are needed by the particular fusion algorithm, the maximum operating range must be limited to that at which all the sensors function.

Usually the selected sensors have different size footprints. The issue then is over which footprint to compare the multisensor target reports. The obvious choice is to pick the largest footprint. That way data are compared over an area corresponding to the limiting or least resolution sensor (assuming the footprint represents one pixel). The finer resolution sensors, such as IR, must then acquire and process imagery over the larger footprint before sending the results onto the fusion processor as in sensor-level fusion.

When sensors are not collocated, algorithms and their associated parameters are defined to align the multiple sensor data in time and space. These spatial alignment algorithms take into account the coordinate systems that measure the location of the objects and the errors introduced by transforming the measurements into other coordinates. Uncertainties in object location reflected in position or velocity error volumes are typically included in the coordinate transformations. Gates are established to control data association from different sensors and from temporal and spatial measurements. The gate size is selected to obtain a balance between maximizing detection probability (use of large-sized gates) and minimizing misassociation probability (use of small-sized gates).

Several researchers have addressed the problem of registration of MMW and IR data.[36,37,38] IR sensors typically provide high-resolution imagery in the elevation and azimuth planes, while MMW sensors provide resolution in range and azimuth. Scene registration is made easier if, in the design process, the sensors' fields of view are made as equal as design, operating, and cost constraints permit. Scene registration is also affected by operational constraints, such as special topology or potential false targets, and test conditions where sensor mounting, boresighting, and data analysis issues are addressed. In registering MMW and IR sensor data in pixel-level fusion applications, for example, flat versus rolling terrain topology must be accounted for as part of the data analysis task in order to get valid results from the data fusion process.

3.8 Summary

Data fusion is modeled through low-level and high-level processes. The low-level processes include target detection, classification, identification, and tracking. High-level processes encompass situation and threat assessment. The algorithms generally used to support target detection, classification, and identification are based on physical models, feature-based inference, and cognition. Numerous examples of these techniques were introduced, including Kalman filtering, Bayesian inference, Dempster-Shafer, generalized evidence processing, artificial neural networks, clustering, voting logic, pattern recognition, and fuzzy set theory. Different algorithms are used for track estimation. The track estimation algorithms are concerned with data alignment, data and object correlation, and position, kinematic, and attribute estimation. Data alignment establishes a common space-time reference for fusion processing. Correlation is performed through prediction gates; kinematic, attribute, and time association metrics; and data and track association techniques. Prediction gates associate data into candidates that are suitable for updating tracks or forming tentative new tracks. Multiple reports (data) can arise from overlapping gates, multiple returns in a gate, clutter, new targets in a gate, and returns received over multiple scans. Metrics quantify the similarity of the observations. In the context of a multiple target and multiple sensor environment, data association applies the metric to compare tracks and measurement data (reports) from different sensors to determine candidates for the correlation process. Correlation determines if the tracks and reports resulting from association belong to a common object. Track association merges tracks from different sensors to form a central track file. Position, kinematic, and attribute estimation combine information from multiple observations to improve knowledge of the target's position, velocity, and identification.

Data fusion that supports situation assessment interprets current relationships among objects and events in the context of an operational environment. Important functions included in situation assessment are object, event, and activity aggregation and contextual interpretation and fusion. Fusion in support of threat assessment is designed to estimate enemy capabilities, threat opportunities, enemy intent, and levels of danger. Included in threat assessment are estimation of enemy capability and intent, identification of threats, multiperspective assessment, and analysis of friendly and enemy capabilities.

Data fusion architectures are described in several ways. The first taxonomy is based on the amount of data processing performed by the sensors, data products produced by the sensors, and the location of the fusion processes. In this case, the architectures are referred to as sensor-level fusion (or autonomous fusion, distributed fusion, and post-individual sensor processing fusion), central-level fusion (or centralized fusion and pre-individual sensor processing fusion), and hybrid fusion (using combinations of the sensor-level and central-level approaches). The second fusion lexicon utilizes the resolution of the data and the extent of the processing performed by a sensor before the data are fused. The nomenclature used in this case is pixel-level, feature-level, and decision-level fusion. Sensor-level fusion allows signal processing to be optimized for the individual sensors in the architecture, while central-level fusion can be designed to optimally process all the data arriving from the entire suite of sensors. Other considerations arise in selecting an appropriate architecture, such as data processing and communication resources and the use to which the fusion products are put.

References

1. F.E. White, Jr., "Joint directors of laboratories data fusion subpanel report: SIGINT session," *Tech. Proc. Joint Service Data Fusion Symposium*, Vol. I, DFS-90, 469-484 (1990).
2. Data Fusion Development Strategy Panel, *Functional Description of the Data Fusion Process*, Office of Naval Technology (November 1991).
3. E. Waltz and J. Llinas, *Multisensor Data Fusion*, Artech House, Norwood, MA (1990).
4. "1989 data fusion survey," 1990 Data Fusion Symposium, Johns Hopkins University Applied Physics Laboratory, Laurel, MD (May 1990).
5. J. Johnson, Paper AD-220160, Image Intensifier Symposium, Fort Belvoir, VA, 49 (Oct. 1958).
6. F.A. Rosell and R.H. Willson, "Recent psychophysical experiments and the display signal-to-noise ratio concept," Chap. 5 in *Perception of Displayed Information*, Editor: L.M. Biberman, Plenum Press, New York, NY (1973).
7. G.A. Gordon, R.L. Hartman, and P.W. Kruse, "Imaging-mode operation of active NMMW systems," Chapter 7 in *Infrared and Millimeter Waves, Vol. 4: Millimeter Systems*, Editor: K.J. Button and J.C. Wiltse, Academic Press, New York, NY (1981).
8. L.A. Klein, *Millimeter-Wave and Infrared Multisensor Design and Signal Processing*, Artech House, Norwood, MA (1997).
9. D.L. Hall and R.J. Linn, "Algorithm selection for data fusion systems," 1987 *Tri-Service Data Fusion Symposium Technical Proceedings*, Vol. I, DFS-87, Naval Air Development Center, Warminster, PA (June 1987).
10. D.L. Hall and R.J. Linn, "A taxonomy of algorithms for multi-sensor data fusion," *Tech. Proc. Joint Service Data Fusion Symposium*, Vol. I, DFS-90, 594-610, Naval Air Development Center, Warminster, PA (1990).
11. D.L. Hall, *Mathematical Techniques in Multisensor Data Fusion*, Artech House, Norwood, MA (1992).
12. I.R. Goodman, R.P.S. Mahler, and H.T. Nguyen, *Mathematics of Data Fusion*, Kluwer Academic Publishers, Norwell, MA (1997).
13. J. Pearl, *Probabilistic Reasoning in Intelligent Systems: Networks of Plausible Inference*, Morgan Kaufmann Publishers, San Mateo, CA (1988).
14. S.C.A. Thomopoulos, "Theories in distributed decision fusion: comparison and generalization," *Sensor Fusion III: 3-D Perception and Recognition, Proc. SPIE* 1383, 623-634 (1990).
15. S.C.A. Thomopoulos, "Sensor integration and data fusion," *J. Robot. Sys.*, 7(3), 337-372 (June 1990).
16. R. Schalkoff, *Pattern Recognition: Statistical, Structural, and Neural Approaches*, John Wiley, New York, NY (1992).

17. S.S. Blackman, "Theoretical approaches to data association and fusion," *Sensor Fusion, Proc. SPIE* 931, 50-55 (1988).
18. J. Munkres, "Algorithms for the assignment and transportation problem," *J. SIAM*, 5, 32-38 (March 1957).
19. O.E. Drummond, D.A. Castanon, and M.S. Bellovin, "Comparison of 2-D assignment algorithms for sparse, rectangular, floating point, and cost matrices," *J. SDI Panels Tracking,* Institute for Defense Analyses, Alexandria, VA, Issue No. 4, 4-81 to 4-97 (Dec. 15, 1990).
20. I.P. Bottlik and S.S. Blackman, "Coordinated presentation of multiple hypotheses in multitarget tracking," *Signal and Data Processing of Small Targets, Proc. SPIE* 1096, 152-159 (April 1989).
21. Y. Bar-Shalom and E. Tse, "Tracking in clutter environments with probabilistic data association," *Automat.*, 11, 451-460 (September 1975).
22. Y. Bar-Shalom and T.E. Fortmann, *Tracking and Data Association*, Academic Press, New York, NY (1988).
23. H.A.P. Bloom and Y. Bar-Shalom, "The interacting multiple-model algorithm for systems with switching coefficients," *IEEE Trans. Automat. Control*, AC-33, 780-783 (August 1988).
24. C.L. Morefield, "Application of 0-1 integer programming to multitarget tracking problem," *IEEE Trans. Automat. Control*, AC-22, 302-312 (June 1977).
25. E. Rechtin, *Systems Architecting: Creating and Building Complex Systems*, Prentice Hall, Englewood Cliffs, NJ, 1991.
26. Draft Military Standard, *Systems Engineering*, MIL-STD-499B (May 15, 1991).
27. Data Fusion Subpanel, *Data Fusion Lexicon*, Joint Directors of Laboratories Technical Panel for C^3 (October 1991).
28. R.C. Luo and M.G. Kay, "Multisensor integration and fusion: issues and approaches," *Sensor Fusion, Proc. SPIE* 931, 42-49 (1988).
29. S.S. Blackman, "Multiple sensor tracking and data fusion," Chapter 7 in S.A. Hovanessian, *Introduction to Sensor Systems*, Artech House, Norwood, MA (1988).
30. V.G. Comparato, "Fusion–the key to tactical mission success," *Sensor Fusion, Proc. SPIE* 931, 2-7 (1988).
31. S.S. Blackman, "Association and Fusion of Multiple Sensor Data," Chapter 6 in *Multitarget-Multisensor Tracking: Advanced Applications*, Editor: Y. Bar-Shalom, Artech House, Norwood, MA (1990).
32. G.S. Robinson and A.O. Aboutalib, "Trade-off analysis of multisensor fusion levels," *Proc. of the 2nd National Symposium on Sensors and Sensor Fusion*, Vol. II, GACIAC PR 89-01, 21-34, IIT Research Institute, Chicago, IL, (1990).
33. S. Ramo, J.R. Whinnery, and T. Van Duzer, *Fields and Waves in Communication Electronics*, John Wiley, New York, NY (1967).

34. A.O. Aboutalib and T.K. Luu, "An efficient target extraction technique for laser radar imagery, *Digital Signal Processing, Association, and Tracking of Point Source, Small, and Cluster Targets*, *Proc. SPIE* 1096 (1989).
35. P.F. Castalez and J.L. Dana, "Neural networks in data fusion applications," *Proc. of the 2nd National Symposium on Sensors and Sensor Fusion,* Vol. II, GACIAC PR 89-01, 105-114, IIT Research Institute, Chicago, IL (1990).
36. K.S. Strom, J.F. Carter, and J.H. Chockley, "Joint Army/Air Force millimeter wave/infrared seeker development," *Proc. of the 4th National Symposium on Sensor Fusion,* Vol. I, 213400-93-X(I), 63-81, ERIM, Ann Arbor, MI (1991).
37. J.E. Overland et al., "Identifying the density of multi-sensor false alarms," *Proc. of the 4th National Symposium on Sensor Fusion,* Vol. I, 213400-93-X(I), 313-322, ERIM , Ann Arbor, MI (1991).
38. E.D. D'Anna and M.A. Richards, "A radar angular resolution improvement technique for multispectral sensor beam registration," *Proc. of the 4th National Symposium on Sensor Fusion,* Vol. I, 213400-93-X(I), 417-429, ERIM, Ann Arbor, MI (1991).

Chapter 4

Bayesian Inference

Bayesian inference is a statistical-based data fusion algorithm based on Bayes' rule. As such, it belongs to the class of algorithms that uses *a priori* knowledge about the observation space to make inferences about the identity of objects in that space. Bayesian inference provides a method for calculating the conditional or *a posteriori* probability of a hypothesis being true given supporting evidence.

4.1 Bayes' rule

Bayes' rule is derived by evaluating the probability of occurrence of an arbitrary event E assuming that another event H has occurred. The probability is given by[1]

$$P(E \mid H) = \frac{P(EH)}{P(H)}, \tag{4-1}$$

where H is an event with positive probability. The quantity $P(E|H)$ is the probability of E conditioned on the occurrence of H. The conditional probability is not defined when H has zero probability.

Equation (4-1) may be rewritten as

$$P(EH) = P(E \mid H)\, P(H), \tag{4-2}$$

which is referred to as the theorem on compound probabilities.

A formula useful when conditional probabilities are easier to evaluate than unconditional probabilities may be derived when H consists of a set of mutually exclusive and exhaustive hypotheses. The mutually exhaustive property implies that one hypothesis necessarily occurs, i.e., the union of H_1, …, H_n is the entire sample space. Under these conditions, any event E can occur only in conjunction with some H_j such that

$$E = EH_1 \cup EH_2 \cup \ldots \cup EH_n\,. \tag{4-3}$$

Since the EH_j are mutually exclusive, their probabilities add as

$$P(E) = \sum_{j=1}^{n} P(E H_j). \tag{4-4}$$

Upon substituting H_j for H and summing over j, Eq. (4-2) becomes

$$P(E) = \sum_j [P(E \mid H_j)\ P(H_j)], \tag{4-5}$$

when the identity in Eq. (4-4) is applied.

In Bayesian inference, we are interested in the probability that hypothesis H_i is true given the existence of evidence E. This statement is expressed as

$$P(H_i \mid E) = \frac{P(EH_i)}{P(E)}. \tag{4-6}$$

If Eqs. (4-2) and (4-5) are introduced into Eq. (4-6), Eq. (4-6) takes the form of Bayes' rule as

$$P(H_i \mid E) = \frac{P(E \mid H_i)\ P(H_i)}{\sum_j [P(E \mid H_j)\ P(H_j)]}, \tag{4-7}$$

where

$P(H_i \mid E)$ = *a posteriori* probability that hypothesis H_i is true given evidence E,

$P(E \mid H_i)$ = probability of observing evidence E given that H_i is true (sometimes referred to as the likelihood function of H_i),

$P(H_i)$ = *a priori* probability that hypothesis H_i is true,

$$\sum_j P(H_j) = 1, \tag{4-8}$$

and

$\sum_j P(E \mid H_j) P(H_j)$ = preposterior or probability of observing evidence E given that hypothesis H_j is true for all hypotheses j.

The likelihood functions represent the extent to which the *a posteriori* probability is subject to change. These functions are evaluated either through offline experiments or by analyzing the information available in the problem at hand.

The preposterior is simply the sum of the products of the likelihood functions and the *a priori* probabilities.

4.2 Comparison of Bayesian inference with classical inference

Bayes' formulation of conditional probability is satisfying for several reasons. First, it provides a determination of the probability of a hypothesis being true, given the evidence. By contrast, classical inference gives the probability that an observation can be attributed to an object or event, given an assumed hypothesis. Second, Bayes' formulation allows incorporation of *a priori* knowledge about the likelihood of a hypothesis being true at all. Third, Bayes permits the use of subjective probabilities for the *a priori* probabilities of hypotheses and for the probability of evidence given a hypothesis when empirical data are not available. This attribute permits a Bayesian inference process to be applied to multisensor fusion since probability density functions are not required. However, the output of such a process is only as good as the input *a priori* probability data.

Bayesian inference therefore resolves some of the difficulties that occur with classical inference methods as shown in Table 4.1. However, Bayesian methods require the *a priori* probabilities and likelihood functions be defined, introduce complexities when multiple hypotheses and multiple conditional dependent events are present, require that competing hypotheses be mutually exclusive, and cannot support an uncertainty class as does Dempster-Shafer.[2,3]

Parzen gives an example that illustrates the impact of *a priori* knowledge about the likelihood of a hypothesis being true on the value of the *a posteriori* probability computed by Bayes' rule.[4] Suppose a patient visits his physician who proceeds to administer a low cost screening test for cancer. The test has an accuracy of 95 percent (i.e., the test will indicate positive 95 percent of the time if the patient has the disease) with a 4 percent false alarm probability. Furthermore, suppose that cancer occurs in 5 out of every 1000 people in the population. If the patient is informed that he has tested positively for cancer, what is the probability he actually has cancer?

The Bayesian formulation predicts the required probability as

$$P(\text{patient has cancer} \mid \text{test positive}) = \frac{P(\text{test positive} \mid \text{given cancer})\, P(\text{cancer})}{P(\text{test positive})}, \tag{4-9}$$

where

$$\begin{aligned} P(\text{test positive}) = {} & P(\text{test positive} \mid \text{given cancer})\, P(\text{cancer}) \\ & + P(\text{test positive} \mid \text{no cancer})\, P(\text{no cancer}). \end{aligned} \tag{4-10}$$

Table 4.1: Comparison of classical and Bayesian inference

Classical	Bayesian
Features of the model:	
Probability model links observed data and a population.	Probability of a hypothesis being true is determined from known evidence.
Probability model is usually empirical based on a large number of samples.	Likelihood of a hypothesis is updated using a previous likelihood estimate and additional evidence.
A number of decision rules may be used to decide between the null hypothesis H_0 and an opposing hypothesis H_1, including maximum likelihood, Neyman-Pearson, and minimax. Other cost functions available for use with Bayesian inference are maximum *a posteriori* and Bayes.[3,5,6] [1]	Either classical probabilities or subjective probability estimates may be used (i.e., probability density functions are not necessarily required).
	Subjective probabilities are inferred from experience and vary from person to person.
	Supports more than two hypotheses at a time.
Disadvantages:	
When generalized to include multi-dimensional data from multiple sensors, *a priori* knowledge and multidimensional probability density functions are required.	*A priori* probabilities and likelihoods must be defined.
Generally, only two hypotheses can be assessed at a time, namely H_0 and H_1.	Complexities are introduced when multiple hypotheses and multiple conditional dependent events are present.
Multivariate data produce evaluation complexities.	Competing hypotheses must be mutually exclusive.
A priori assessments are not utilized.	Cannot support an uncertainty class.

[1]. *Maximum likelihood:* Accepts hypothesis H_0 as true if the probability $P(H_0)$ of H_0 multiplied by $P(y \mid H_0)$ is greater than $P(H_1) \times P(y \mid H_1)$.

Neyman-Pearson: Accepts the hypothesis H_0 if the ratio of the likelihood function for H_0 to the likelihood function for H_1 is less than or equal to a constant c. The constant is selected to give the desired significance level.

Minimax: A cost function is constructed to quantify the risk or loss associated with choosing a hypothesis or its alternative. The minimax approach selects H_0 such that the maximum possible value of the cost function is minimized.

Maximum a posteriori: Accepts hypothesis H_0 as true if the probability $P(H_0 \mid y)$ of H_0 given observation y is greater than the probability $P(H_1 \mid y)$ of H_1 given observation y.

Bayes: A cost function is constructed that provides a measure of the consequences of choosing hypothesis H_0 versus H_1. This decision rule selects the hypothesis that minimizes a cost function based on detection and false alarm probabilities.

The probability P(test positive|no cancer) is the false alarm probability or Type 1 error. The Type 2 error is the probability of missing the detection of cancer in a patient with the disease.

Upon substituting the statistics for this example into Eq. (4-9), we find

$$P(\text{patient has cancer}|\text{test positive}) = \frac{(0.95)(0.005)}{(0.95)(0.005) + (0.04)(0.995)} = 0.107 \quad (4\text{-}11)$$

or 10.7 percent. Intuitively, this result appears to be smaller than expected. It asserts that in only 10.7 percent of the cases in which the test gives a positive result and declares cancer to be present is it actually true that cancer is present. Further testing is thus required when this type of initial test is administered. The screening test may be said to be reliable because it will detect cancer in 95 percent of the cases in which cancer is present. However, the critical Type 2 error is 0.05, implying that the test will not diagnose 1 in 20 cancers.

To increase the probability of the patient actually having cancer, given a positive test, and concurrently reduce the Type 2 error requires a test with a greater accuracy. A more effective method of increasing the *a posteriori* probability is to reduce the false alarm probability. If, for example, the test accuracy is increased to 99.9 percent and the false alarm probability reduced to 1 percent, the probability of the patient actually having cancer, given a positive test, is increased to 33.4 percent. The Type 2 error now implies a missed diagnosis in only 1 out of 1,000 patients. Increasing the test accuracy to 99.99 percent has a minor effect on the *a posteriori* probability, but reduces the Type 2 error by another order of magnitude.

In other situations, the Type 1 error may be the more serious error. Such a case would occur if an innocent man was being tried for a crime and his sentence relied on the outcome of a certain experiment. If a hypothesis corresponding to his innocence was constructed and was rejected by the experiment, then an innocent man would be convicted and a Type 1 error would result. On the other hand, if the man was guilty and the experiment accepted the hypothesis corresponding to innocence, the guilty man would be freed and a Type 2 error would result.[7]

When Bayesian inference is used to support the fusion of identity information, the process may be described as illustrated in Figure 4.1. Here multiple sensors are shown observing parametric data [e.g., IR signatures, radar cross section, pulse repetition interval, and identification-friend-or-foe (IFF)] about an entity whose identity is unknown. Each of the sensors provides an identity declaration D or hypothesis about the object's identity based on the observations and a sensor-specific algorithm. The previously established, i.e., *a priori*, performance characteristics of each sensor's classification algorithm (developed either

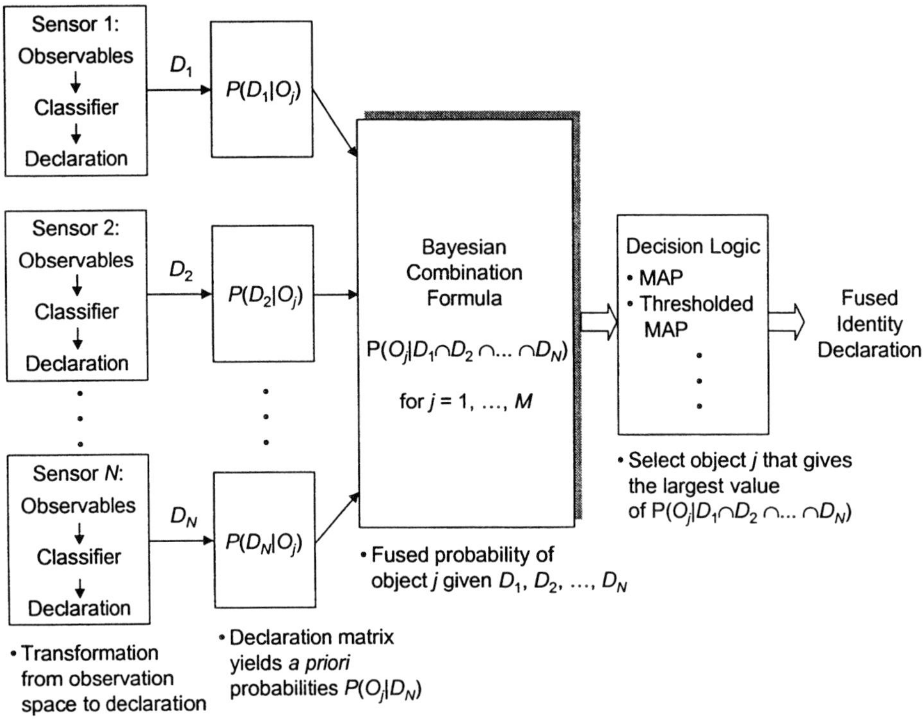

Figure 4.1: Bayesian fusion process
(Adapted from E. Waltz and J. Llinas, *Multisensor Data Fusion*, Artech House, Norwood, MA [1990])

theoretically or experimentally) provide estimates of the probability that the sensor will declare the object to be a certain type, given that the object is in fact type j. This probability is denoted by $P(D \mid O_j)$. These declarations are then combined using a generalization of Eq. (4-1) to produce an updated, joint probability for each entity O_j based on the multisensor declarations.

Thus, the probability of having observed object j from the set of M objects given declaration (evidence) D_1 from Sensor 1, declaration D_2 from Sensor 2, etc., is

$$P(O_j \mid D_1 \cap D_2 \cap D_3 \cap \ldots \cap D_N), j = 1, \ldots, M. \tag{4-12}$$

By applying a decision logic, a joint declaration of identity can be selected by choosing the object whose joint probability given by Eq. (4-12) is greatest. The choice of the maximum value of Eq. (4-12) is referred to as the maximum *a posteriori* probability (MAP) decision rule. Other decision rules exist as indicated in Figure 4.1 and Table 4.1. The Bayes formulation, therefore, provides a method to combine identity declarations from multiple sensors to obtain a new and hopefully improved joint identity declaration. Required inputs for the Bayes method are the ability to compute or model $P(E \mid H_i)$ for each sensor and entity and the *a priori* probabilities that the hypotheses $P(H_i)$ are true. When *a priori* information is lacking concerning the relative likelihood of H_i, the principle of indifference may be used in which $P(H_i)$ for all i are set equal.

The application of Bayes' rule is often contrasted in modern probability theory with the application of confidence intervals.[8] While Bayes' rule provides an inference approach suitable for some data fusion applications, the theory of confidence intervals is better suited when it is desired to assert, with some specified probability, that the true value of a certain parameter (e.g., mean and variance) that characterizes a known distribution is situated between two limits.

4.3 Enhancing underground mine detection with data from two non-commensurate sensors

The detection of buried mines may be enhanced by fusing data from multiple sensors that respond to signatures that are generated by independent phenomena. Two sensors that meet this criterion are metal detectors and ground penetrating radars. The metal detector (MD) indicates the presence of metal fragments larger than 1 cm with weight exceeding a few grams. The ground penetrating radar (GPR) detects objects larger than approximately 10 cm that differ in electromagnetic properties from the soil or background material. While the metal detector simply distinguishes between objects that contain or do not contain metal, the GPR supports object classification since it is responsive to multiple characteristics of the object such as size, shape, material type, and internal design.

In an experiment reported by Brusmark et al., a low metal content mine, metal fragments, plastic, beeswax (an explosive simulant), and stone were buried in sand at a 5-cm depth.[9] The metal detector provided a signal whose amplitude was proportional to the metal content of the object. The GPR transmitted a broadband waveform covering 300 to 3000 MHz. The antenna footprint consisted of four separate lobes, with a common envelope of about 30 cm. An artificial neural network was trained to classify the buried objects that were detected by the GPR. The inputs to the neural network were features produced by Fourier transform analysis, bispectrum transform analysis, wavelet transform analysis, and local frequency analysis of the GPR signals.

Bayesian inference was used to compute the *a posteriori* probabilities that the detected object belonged to one of the object classes represented by mine (MINE), not mine ($\overline{\text{MINE}}$), or background (BACK). The Bayesian approach to data fusion is based on updating probabilities as illustrated in the influence diagram shown in Figure 4.2.

When previous and current period values of the *a posteriori* probability are required, as in this example, Bayes' rule is written as[2]

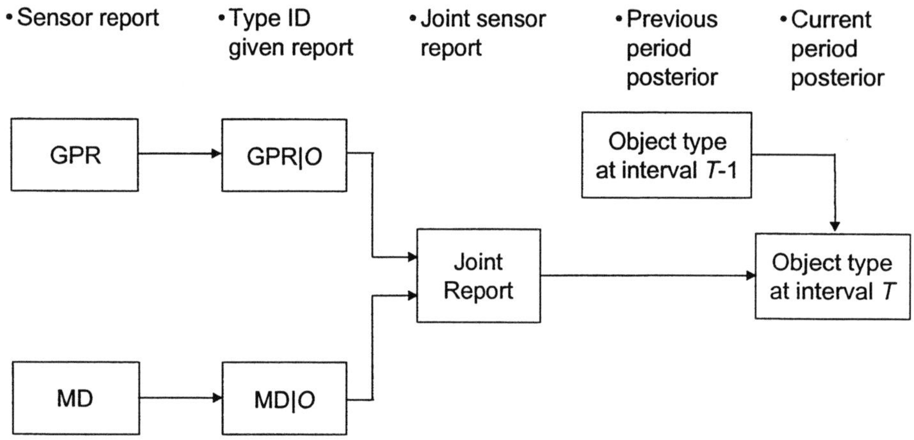

Figure 4.2: Influence diagram for two-sensor mine detection

$$P(C_i \mid BA) = \frac{P(B \mid C_i A)\ P(C_i \mid A)}{\sum_j [P(B \mid C_j A)\ P(C_j A)]}, \tag{4-13}$$

where

$P(C_i \mid BA)$ = *a posteriori* probability of C_i given the data B at the prior state A,

$P(B \mid C_i A)$ = probability of observing data B given C_i as computed using the information at the prior state A (i.e., the likelihood function of C_i),

$P(C_i \mid A)$ = *a priori* probability of C_i based on the information at the prior state A, and

$\sum_j P(B \mid C_j A) P(C_j \mid A)$ = preposterior or probability of the data B occurring given the prior state of information, but conditioned on all possible outcomes C_j.

The probability of the sensors observing data conditioned on object type is given by

$$\begin{aligned} P_{MD}(\text{data} \mid O_j) = {} & P_{MD}(\text{data} \mid \text{MINE})\ P(\text{MINE} \mid O_j) \\ & + P_{MD}(\text{data} \mid \overline{\text{MINE}})\ P(\overline{\text{MINE}} \mid O_j) \\ & + P_{MD}(\text{data} \mid \text{BACK})\ P(\text{BACK} \mid O_j) \end{aligned} \tag{4-14}$$

and

$$P_{GPR}(\text{data} \mid O_j) = P_{GPR}(\text{data} \mid \text{MINE})\ P(\text{MINE} \mid O_j) + P_{GPR}(\text{data} \mid \overline{\text{MINE}})\ P(\overline{\text{MINE}} \mid O_j) + P_{GPR}(\text{data} \mid \text{BACK})\ P(\text{BACK} \mid O_j) \tag{4-15}$$

where MD denotes the mine sensor, GPR the ground penetrating radar, and O_j an object of type *j*. The set of arrows from "sensor report" to "type identification given report" in Figure 4.2 represents the probability calculations defined by Eqs. (4-14) and (4-15).

The values of the likelihood functions for the metal detector, namely P_{MD}(data | MINE), $P_{MD}(\text{data} \mid \overline{\text{MINE}})$, and P_{MD}(data | BACK), and for the ground penetrating radar, namely, P_{GPR}(data | MINE), $P_{GPR}(\text{data} \mid \overline{\text{MINE}})$, and P_{GPR}(data | BACK), are found through *a priori* measurements. The mine detector "data" are equal to the preprocessed signal amplitude, and $P_{MD}(\text{data} \mid O_j)$ is equal to the probability of receiving a signal of some amplitude given the object is of type O_j. These probabilities are found from extensive experiments with buried mine-like objects consisting of different materials and sizes. The ground penetrating radar signal profile data in the scanned area are input to an artificial neural network trained to identify antipersonnel mines. The output of the neural network over many experiments gives $P_{GPR}(\text{data} \mid O_j)$. Quantitative values for $P(\text{MINE} \mid O_j)$, $P(\overline{\text{MINE}} \mid O_j)$, and $P(\text{BACK} \mid O_j)$ are dependent on the types and numbers of objects in the mine infected area.

Next, the joint sensor report shown in Figure 4.2 is computed for a given time interval as the product of Eqs. (4-14) and (4-15) since the sensors respond to signatures generated by independent phenomena, i.e., they are noncommensurate. Thus, the joint probability of detection is

$$P(\text{data} \mid O_j) = \prod_i P_i(\text{data} \mid O_j), \tag{4-16}$$

where *i* is the sensor index, here equal to 1 and 2.

Finally, Bayes' rule is applied to calculate the *a posteriori* probability $P(O_j \mid \text{data})$ that the detected object is of type *j* based on the value of $P(O_j \mid \text{data})$ evaluated in the previous period and the *a priori* probabilities. Accordingly, from Eq. (4-13),

$$P(O_j \mid \text{data}) = \frac{P(\text{data} \mid O_j) P(O_j)}{P(\text{data})}, \tag{4-17}$$

where

$P(O_j)$ = value of $P(O_j \mid \text{data})$ during the previous period and

$$P(\text{data}) = \sum_j P(\text{data} \mid O_j)\, P(O_j) \tag{4-18}$$

is the preposterior or probability of observing the data collected during the previous period given that objects O_j are present. Larger values of P(data) imply that the previous period values are more predictive of the situation as it evolves. When the sensors do not report an object type for the current time interval, updating is not performed and the values of $P(O_j|\text{data})$ for the current interval are set equal to those from the previous period.

Since the primary task in this example is to locate mines, the second and third terms in Eqs. (4-14) and (4-15) are combined into a single declaration $\overline{\text{MINE}}$ that represents the absence of a mine. The problem is further simplified by choosing O_1 = MINE (in this experiment, the mine was an antipersonnel mine) and $O_2 = \overline{O_1}$. Therefore, the required probabilities are only dependent on $P(\text{data} \mid O_1)$ since

$$P(\text{data} \mid O_2) = 1 - P(\text{data} \mid O_1) \tag{4-19}$$

and

$$P(O_2) = 1 - P(O_1). \tag{4-20}$$

An initial value for $P(O_1)$ and lower and upper bounds inside the interval (0, 1) for admissible values of $P(O_1|\text{data})$ are needed to evaluate Eq. (4-17). Since 5 different types of objects were buried, $P(O_1)$ was initially set equal to 1/5. The boundaries for $P(O_1|\text{data})$ were limited to (0.01, 0.99) to prevent the process that computes the *a posteriori* probability from terminating prematurely at the limiting endpoint values of 0 and 1. The updated joint probability of detection by the sensors is found by applying Eq. (4-17) to the joint MD and GPR reports as represented by a matrix formed by the scanned data. This process is enhanced by passing the GPR signatures through a matched filter to remove the distortion caused by the antenna pattern.[10]

The posterior probabilities for object classes mine, not mine, and background are computed from the posterior probabilities for object type and the scenario defined values for $P(\text{MINE} \mid O_j)$, etc., as

$$P(\text{MINE} \mid \text{data}) = \sum_j [P(O_j \mid \text{data})\ P(\text{MINE} \mid O_j)], \tag{4-21}$$

$$P(\overline{\text{MINE}} \mid \text{data}) = \sum_j [P(O_j \mid \text{data})\ P(\overline{\text{MINE}} \mid O_j)], \tag{4-22}$$

and

$$P(\text{BACK} \mid \text{data}) = \sum_j [P(O_j \mid \text{data})\ P(\text{BACK} \mid O_j)]. \tag{4-23}$$

Thus, the probability of locating a mine is the sum of individual probabilities that are dependent on the identification of various features. The term $P(\text{MINE} \mid O_j)$ expresses the *a priori* probability of finding a mine conditioned on object type O_j being present. In this particular application where metal detector and ground penetrating radar data were fused, it was assumed that very low metal content mines could be detected by the metal detector alone. Two cautions were mentioned by the authors, however. The first was that the data fusion algorithms should be robust in their ability to identify objects other than those expected to be found. Second, since metal may not be detected by the metal detector, the multisensor system must be designed to rely on ground penetrating radar detections alone to identify objects.

4.4 Summary

Bayes' rule has been derived from the classical expression for the conditional probability of the occurrence of an event given supporting evidence. Bayes' formulation of conditional probability is satisfying for several reasons. First, it provides a method to compute the probability of a hypothesis being true, given the evidence. Second, Bayes' formulation allows incorporation of *a priori* knowledge about the likelihood of a hypothesis being true at all. Third, Bayes permits the use of subjective probabilities for the *a priori* probabilities of hypotheses and for the probability of evidence given a hypothesis. This attribute permits Bayesian inference to be applied to multisensor fusion since probability density functions are not required. However, the output of such a process is only as good as the input *a priori* probability data. An application of Bayesian inference was presented to illustrate how data gathered by multiple sensors could be combined to enhance the detection of buried mines. The example also described how the *a posteriori* probabilities were updated for the current time period using values calculated in the previous period.

References

1. W. Feller, *An Introduction to Probability Theory and its Applications*, 2nd Ed., John Wiley and Sons, New York, NY (1962).
2. E. Waltz and J. Llinas, *Multisensor Data Fusion*, Artech House, Norwood, MA (1990).
3. D.L. Hall, *Mathematical Techniques in Multisensor Data Fusion*, Artech House, Norwood, MA (1992).
4. E. Parzen, *Modern Probability Theory and Its Applications*, John Wiley and Sons, New York, NY (1964).
5. G.S. Robinson and A.O. Aboutalib, "Trade-off analysis of multisensor fusion levels," *Proc. of the 2nd National Symposium on Sensors and Sensor Fusion*, Vol. II, GACIAC PR 89-01, 21-34, IIT Research Institute, Chicago, IL, (1990).
6. D.C. Lai and R.D. McCoy, "Optimal classification fusion in multi-sensor target recognition systems," *Proc. 2nd National Symposium on Sensors and Sensor Fusion*, Vol. II, GACIAC PR89-01, 259-266, IIT Research Institute, Chicago, IL (1990).
7. P.G. Hoel, *Introduction to Mathematical Statistics*, John Wiley and Sons, New York, NY (1947).
8. H. Cramér, *Mathematical Methods of Statistics*, Princeton University Press, Princeton, NJ (1961).
9. B. Brusmark, A.-L. Christiansen, P. Jägerbro, and A. Lauberts, "Combination of ground penetrating radar and metal detector data for mine detection," *Proc. of 7th International Conference on Ground Penetrating Radar* (*GPR98*), Radar Systems and Remote Sensing Laboratory, Univ. of Kansas, Lawrence, KS, 1, 331-336 (June 1998).
10. P. Jägerbro, B. Brusmark, A.-L. Christiansen, and A. Lauberts, "Combination of GPR and metal detector for mine detection," *IEE 2nd International Conference on the Detection of Abandoned Land Mines*, Edinburgh, 177-181 (1998).

Chapter 5

Dempster-Shafer Algorithm

Dempster-Shafer inference, a statistical-based data fusion classification algorithm, is used when the sensors (or more generally, the information sources) contributing information cannot associate a 100 percent probability of certainty to their output decisions. The algorithm provides a technique that captures and combines whatever certainty exists in the object discrimination capability of the sensors. Knowledge from multiple sensors about events (called propositions) are combined using Dempster's rule to find the intersection or conjunction of the propositions and the associated probability.

5.1 Overview of the process

An overview of the Dempster-Shafer data fusion process is shown in Figure 5.1. Each sensor has a set of observables corresponding to the phenomena that generate information received about objects and their surroundings. In this illustration, a sensor operates on the observables with its particular set of classification algorithms (sensor-level fusion). The knowledge level gathered by each sensor k, where $k = 1, \ldots, N$, associates a declaration of target or object type (referred to in the figure by object o_i where $i = 1, \ldots, n$) with a probability mass $m_k(o_i)$ between 0 and 1. The probability mass expresses the certainty of the decision. Probability masses closer to unity characterize decisions made with more definite knowledge or less uncertainty about the nature of the target. The probability masses for the decisions made by each sensor are then combined using Dempster's rules of combination. The hypothesis favored by the largest accumulation of evidence from all contributing sensors is selected as the most probable outcome of the fusion process. In addition to real-time sensor data, other information or rules can be stored in the information base to improve the overall decision or target discrimination capability. Examples of such rules are "Ships detected in known shipping lanes are cargo vessels" and "Objects in previously charted Earth orbits are reconnaissance or weather satellites."

A computer stores the relevant information from each sensor. The converse is also true, that is, targets not supported by evidence from any sensor are not stored. The data fusion process consists of finding the intersection of two or more sensor outputs and any applicable rules.

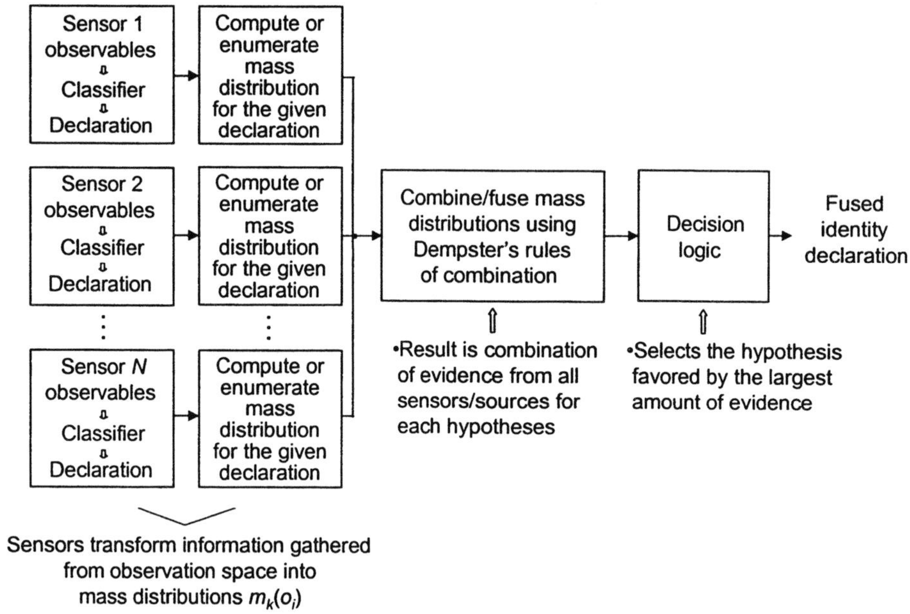

Figure 5.1: Dempster-Shafer data fusion process
(Adapted from E. Waltz and J. Llinas, *Multisensor Data Fusion*, Artech House, Norwood, MA [1990])

5.2 Implementation of the method

Assume a set of n mutually exclusive and exhaustive propositions exists, for example, a target is of type $a_1, a_2, \dots$, or a_n. This is the set of all propositions making up the hypothesis space, called the frame of discernment, and is denoted by Θ. A probability mass $m(a_i)$ is assigned to any of the original propositions or to the union of the propositions based on available sensor information. Thus, the union or disjunction that the target is of type a_1 or a_2 (denoted $a_1 \cup a_2$) has probability mass assignment $m(a_1 \cup a_2)$. The number of propositions (including all possible unions, but excluding the null set) that exists is called the power set and is equal to $2^n - 1$. For example if $n = 3$, there are $2^3 - 1 = 7$ propositions given by $a_1, a_2, a_3, a_1 \cup a_2, a_1 \cup a_3, a_2 \cup a_3$, and $a_1 \cup a_2 \cup a_3$.

In the event that all the probability mass cannot be directly assigned by the sensor to any of the propositions or their unions, the remaining mass is assigned to the frame of discernment Θ (representing uncertainty as to further definitive assignment) as $m(\Theta) = m(a_1 \cup a_2 \cup \dots \cup a_n)$ or to the negation of a proposition such as $m(\bar{a}_1) = m\,(a_2 \cup a_3 \cup \dots \cup a_n)$. A raised bar is used to denote the negation of a proposition. The mass assigned to Θ represents the uncertainty the sensor has concerning the accuracy and interpretation of the evidence.[1] The sum of probability masses over all propositions, uncertainty, and negation equals unity.

To illustrate these concepts suppose that two sensors observe a scene in which there are three targets. Sensor *A* identifies the target as belonging to one of the three possible types a_1, a_2, or a_3. Sensor *B* declares the target to be of type a_1 with a certainty of 80 percent. The intersection of the data from the two sensors is written as

$$(a_1 \text{ or } a_2 \text{ or } a_3) \text{ and } (a_1) = (a_1), \tag{5-1a}$$

or upon rewriting as

$$(a_1 \cup a_2 \cup a_3) \cap (a_1) = (a_1). \tag{5-1b}$$

Only a probability of 0.8 can be assigned to the intersection of the sensor data based on the 80 percent confidence associated with the output from Sensor *B*. The remaining probability of 0.2 is assigned to uncertainty represented by the union (disjunction) of (a_1 or a_2 or a_3).[2]

5.3 Support, plausibility, and uncertainty interval

According to Shafer, "an adequate summary of the impact of the evidence on a particular proposition a_i must include at least two items of information: a report on how well a_i is supported and a report on how well its negation $\overline{a}_i$ is supported."[3] These two items of information are conveyed by the proposition's degree of support and its degree of plausibility.

Support for a given proposition is defined as "The sum of all masses assigned directly by the sensor to that proposition." Thus, the support for target type a_1, denoted by $S(a_1)$, contributed by a sensor is equal to

$$S(a_1) = m(a_1). \tag{5-2}$$

Support for the proposition that the target is either type a_1, a_2, or a_3 is

$$S(a_1 \cup a_2 \cup a_3) = m(a_1) + m(a_2) + m(a_3) + m(a_1 \cup a_2) + m(a_1 \cup a_3) + m(a_2 \cup a_3) + m(a_1 \cup a_2 \cup a_3). \tag{5-3}$$

Plausibility of a given proposition is defined as "The sum of all mass not assigned to its negation." In other words, plausibility defines the mass free to move to the support of a proposition. The plausibility of a_i, denoted by $Pl(a_i)$, is written as

$$Pl(a_i) = 1 - S(\overline{a}_i), \tag{5-4}$$

where $S(\overline{a}_i)$ is called the dubiety and represents the degree to which the evidence impugns a proposition, i.e., supports the negation of the proposition.

Plausibility can also be computed by summing all masses associated with a_i and all unions, including Θ, which contain a_i as

$$Pl(a_i) = m(a_i) + m(a_i \cup a_2) + \ldots + m(\Theta). \tag{5-5}$$

An uncertainty interval is defined by $[S(a_i), Pl(a_i)]$, where

$$S(a_i) \leq Pl(a_i). \tag{5-6}$$

The Dempster-Shafer uncertainty interval shown in Figure 5.2 illustrates the concepts just discussed.[4,5] The lower bound or support for a proposition is equal to the minimal commitment for the proposition based on direct sensor evidence. The upper bound or plausibility is equal to the support plus any potential commitment. Therefore, these bounds show what proportion of evidence is truly in support of a proposition, and what proportion results merely from ignorance or the requirement to normalize the sum of the probability masses to unity.

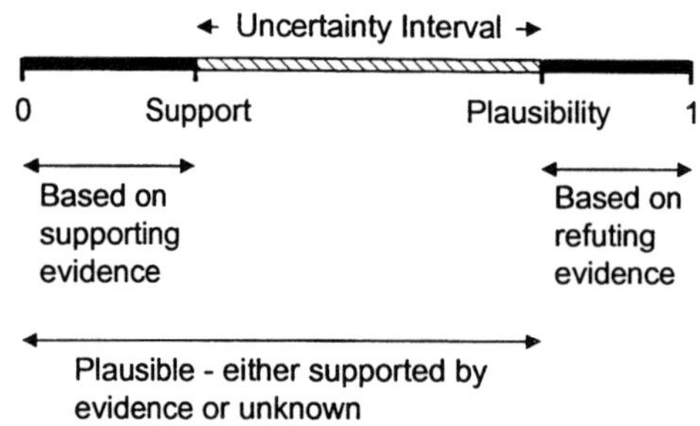

Figure 5.2: Dempster-Shafer uncertainty interval for a proposition

Support and probability mass obtained from a sensor (knowledge source) represent different concepts. Support is calculated from the sum of the probability masses that directly support the proposition and its unions. Probability mass is determined from the sensor's ability to assign some certainty to a proposition based on the evidence.

Table 5.1 provides further interpretations of uncertainty intervals. For example, the uncertainty interval [0,1] represents total ignorance about proposition a_i since there is no direct support for a_i, but also no refuting evidence. The plausible range is equal to unity, as is the uncertainty interval. The uncertainty interval denoted by [0.6,0.6] contains equal support and plausibility values. It indicates a definite probability of 0.6 for proposition a_i since both the direct support and plausibility are 0.6. In this case, the uncertainty interval equals zero. Support and plausibility values represented by [0,0] indicate that the proposition a_i is false as all the probability mass is assigned to the negation of a_i. Therefore, the support for a_i is zero and the plausibility, $1 - S(\overline{a}_i)$, is also zero since $S(\overline{a}_i) = 1$. When a_i is known to be true, [1,1] represents the support and plausibility values.

The uncertainty interval is zero since all the probability mass is assigned to the proposition a_i. Therefore, the support for a_i is 1 and the plausibility, $1 - S(\bar{a}_i)$, is also 1 since $S(\bar{a}_i) = 0$. The support and plausibility values [0.25,1] imply evidence that partially supports proposition a_i with a support value of 0.25. The plausibility of one indicates there is not any direct evidence to refute a_i. All the probability mass in the uncertainty interval of length 0.75 is free to move to the support of a_i. The interval [0,0.85] implies partial support for the negation of a_i since there is no direct evidence to support a_i while there is partial evidence to support $\bar{a}_i$. The support and plausibility represented by [0.25,0.85] show partial direct support for a_i and partial direct support for its negation. In this case, the uncertainty interval represents probability mass that is available to move to support a_i or $\bar{a}_i$.

Table 5.1: Interpretation of uncertainty intervals for proposition a_i

Uncertainty Interval $[S(a_i), Pl(a_i)]$	Interpretation
[0,1]	Total ignorance about proposition a_i
[0.6,0.6]	A definite probability of 0.6 for proposition a_i
[0,0]	Proposition a_i is false
[1,1]	Proposition a_i is true
[0.25,1]	Evidence provides partial support for proposition a_i
[0,0.85]	Evidence provides partial support for $\bar{a}_i$
[0.25,0.85]	Probability of a_i is between 0.25 and 0.85, i.e., the evidence simultaneously provides support for both a_i and $\bar{a}_i$

As an example of how the uncertainty interval is computed from the knowledge a sensor provides, consider once more three targets a_1, a_2, and a_3 observed this time by a single sensor denoted by Sensor A. The frame of discernment Θ is given by

$$\Theta = \{a_1, a_2, a_3\}. \tag{5-7}$$

The negation of proposition a_1 is represented by

$$\bar{a}_1 = \{a_2, a_3\}. \tag{5-8}$$

Assume the probability masses contributed by Sensor A to the propositions a_1, $\bar{a}_1$, $a_1 \cup a_2$, and Θ are

$$m_A(a_1, \bar{a}_1, a_1 \cup a_2, \Theta) = (0.4, 0.2, 0.3, 0.1). \tag{5-9}$$

Using these numerical values, the uncertainty intervals for a_1, $\bar{a}_1$, $a_1 \cup a_2$, and Θ are calculated as shown in Table 5.2. The uncertainty interval calculations for a_1 and $\bar{a}_1$ are straightforward since they are based on direct sensor evidence. The uncertainty interval for proposition $a_1 \cup a_2$ is found using the direct evidence from Sensor A that supports a_1 and $a_1 \cup a_2$. The probability mass $m_1(\Theta)$, i.e., the mass not assignable to a smaller set of propositions, is not included in any of the supporting or refuting evidence for $a_1 \cup a_2$ because $m_1(\Theta)$ represents the residual uncertainty of the sensor in distributing the remaining probability mass directly to any other propositions or unions in Θ based on the evidence. That is, the evidence has allowed the sensor to assign direct probability mass only to propositions a_1, $\bar{a}_1$, and $a_1 \cup a_2$. The remaining mass is assigned to $m_1(\Theta)$, implying that it is distributed in some unknown manner among the totality of all propositions. The uncertainty interval for the proposition Θ is found as follows: Support for Θ is equal to unity because Θ is the totality of all propositions. Plausibility for Θ is also unity because support is not assigned outside of Θ. Therefore, $m_1(\bar{\Theta}) = 0$ and $Pl(\Theta) = 1 - S(\bar{\Theta}) = 1 - 0 = 1$.

Table 5.2: Uncertainty interval calculation for propositions a_1, $\bar{a}_1$, $a_1 \cup a_2$, Θ

Proposition	Support $S(a_i)$	Plausibility $1 - S(\bar{a}_i)$	Uncertainty Interval
a_1	0.4 (given)	$1 - S(\bar{a}_1)$ $= 1 - 0.2 = 0.8$	[0.4, 0.8]
$\bar{a}_1$	0.2 (given)	$1 - S(a_1)$ $= 1 - 0.4 = 0.6$	[0.2, 0.6]
$a_1 \cup a_2$	$S(a_1) + S(a_1 \cup a_2) =$ $0.4 + 0.3 = 0.7$	$1 - S(\overline{a_1 \cup a_2})$ $= 1 - S(\bar{a}_1 \cap \bar{a}_2)$ $= 1 - 0 = 1$*	[0.7, 1]
Θ	$S(\Theta) = 1$	$1 - S(\bar{\Theta})$ $= 1 - 0 = 1$	[1, 1]

*Only probability mass assigned directly by Sensor A to $\bar{a}_1 \cap \bar{a}_2$ is used in the calculation. Since Sensor A has not assigned any probability mass directly to $\bar{a}_1 \cap \bar{a}_2$, the support for $\bar{a}_1 \cap \bar{a}_2$ is zero. Thus, the plausibility of $a_1 \cup a_2$ is unity.

Two parallel sets of terminology, subjective and evidential, are used in the literature to describe the impact of evidence on a proposition. The evidential terminology was used by Shafer to describe the subclass of belief functions that represent evidence. These are the support functions shown in Table 5.3. Other terms in the subjective and evidential vocabulary are also listed.

Table 5.3: Subjective and evidential vocabulary

Subjective	**Evidential**
Belief $Bel(a_i)$	Support $S(a_i)$
Doubt $Dou(a_i) = Bel(\bar{a}_i)$	Dubiety $Dub(a_i) = S(\bar{a}_i)$
Upper Probability $P^*(a_i) = 1 - Bel(\bar{a}_i)$	Plausibility $Pl(a_i) = 1 - S(\bar{a}_i)$

5.4 Dempster's rule for combination of multiple sensor data

Dempster's rule provides the formalism to combine the probability masses provided by multiple sensors or knowledge sources for compatible propositions. The intersection of the propositions having the largest probability mass is selected as the output of the fusion process. Propositions are compatible when their intersection exists. If the intersections of propositions form the null set, the probability masses associated with those intersections are set equal to zero and the probability masses of the nonempty set intersections are increased by a proportionality factor K such that their sum is unity.

The general form of Dempster's rule for the total probability mass committed to an event c is given by

$$m(c) = K \sum_{a_i \cap b_j = c} [m_A(a_i) m_B(b_j)], \tag{5-10}$$

where

$$K^{-1} = 1 - \sum_{a_i \cap b_j = \phi} [m_A(a_i) m_B(b_j)] \tag{5-11}$$

and ϕ is defined as the empty set. If K^{-1} is zero, then m_A and m_B are totally contradictory and the sum defined by Dempster's rule does not exist.

Dempster's rule is illustrated with the following four-target, two-sensor example.

Suppose that four targets are present:

a_1 = friendly target type 1 a_3 = enemy target type 1

a_2 = friendly target type 2 a_4 = enemy target type 2.

The probability mass matrix for target identification contributed by Sensor A is given by

$$m_A = \begin{bmatrix} m_A(a_1 \cup a_3) = 0.6 \\ m_A(\Theta) = 0.4 \end{bmatrix}, \tag{5-12}$$

where $m_A(\Theta)$ is the uncertainty associated with rules used to determine that the target is of type 1.

The probability mass matrix for target identification contributed by Sensor B is given by

$$m_B = \begin{bmatrix} m_B(a_3 \cup a_4) = 0.7 \\ m_B(\Theta) = 0.3 \end{bmatrix}, \tag{5-13}$$

where $m_B(\Theta)$ is the uncertainty associated with the rules used to determine that the target belongs to the enemy.

Dempster's rule is implemented by forming a matrix with the probability masses that are to be combined entered along the first column and last row. Table 5.4 shows the matrix for our example.

Table 5.4: Application of Dempster's rule

First column			
$m_A(\Theta) = 0.4$	$m(a_3 \cup a_4) = 0.28$	$m(\Theta) = 0.12$	
$m_A(a_1 \cup a_3) = 0.6$	$m(a_3) = 0.42$	$m(a_1 \cup a_3) = 0.18$	
	$m_B(a_3 \cup a_4) = 0.7$	$m_B(\Theta) = 0.3$	*Last row*

Matrix (row, column) elements are computed as the product of the probability mass in the same row of the first column and the same column of the last row. The proposition corresponding to a matrix element is equal to the intersection of the propositions that are multiplied. Accordingly, matrix element (1,2) represents the proposition formed by the intersection of uncertainty (Θ) from Sensor A and ($a_3 \cup a_4$) from Sensor B, namely, that the target is enemy type 1 or type 2. The probability mass $m(a_3 \cup a_4)$ associated with the intersection of the propositions is

$$m(a_3 \cup a_4) = m_A(\Theta)\, m_B(a_3 \cup a_4) = (0.4)\,(0.7) = 0.28. \tag{5-14}$$

Matrix element (1,3) represents the intersection of the uncertainty propositions from Sensor A and Sensor B. The probability mass $m(\Theta)$ associated with the uncertainty intersection is

$$m(\Theta) = m_A(\Theta)\, m_B(\Theta) = (0.4)\,(0.3) = 0.12. \tag{5-15}$$

Matrix element (2,2) represents the proposition formed by the intersection of $(a_1 \cup a_3)$ from Sensor A and $(a_3 \cup a_4)$ from Sensor B, namely, that the target is enemy type 1. The probability mass $m(a_3)$ associated with the intersection of these propositions is

$$m(a_3) = m_A(a_1 \cup a_3)\, m_B(a_3 \cup a_4) = (0.6)\,(0.7) = 0.42. \tag{5-16}$$

The proposition represented by $m(a_3)$ has the highest probability mass in the matrix. Thus, it is typically the one selected as the output to represent the fusion of the evidence from Sensors A and B.

Matrix element (2,3) represents the proposition formed by the intersection of $(a_1 \cup a_3)$ from Sensor A and (Θ) from Sensor B. It is calculated as

$$m(a_1 \cup a_3) = m_A(a_1 \cup a_3)\, m_B(\Theta) = (0.6)\,(0.3) = 0.18. \tag{5-17}$$

This probability mass is ascribed to the proposition that the target is type 1, either friendly or hostile.

As an example of how the probability mass of the nonzero elements is increased, suppose that Sensor B had identified targets 2 and 4, instead of targets 3 and 4, with the probability mass assignments given by m_B' as

$$m_B' = \begin{bmatrix} m_B'(a_2 \cup a_4) = 0.5 \\ m_B'(\Theta) = 0.5 \end{bmatrix}. \tag{5-18}$$

Applying Dempster's rule, we find the results shown in Table 5.5 where element (2,2) now belongs to the empty set. Since mass is assigned to ϕ, we find the value for K that redistributes this mass to the nonempty set members.

Table 5.5: Application of Dempster's rule with an empty set

$m_A(\Theta) = 0.4$	$m(a_2 \cup a_4) = 0.20$	$m(\Theta) = 0.20$
$m_A(a_1 \cup a_3) = 0.6$	$m(\phi) = 0.30$	$m(a_1 \cup a_3) = 0.30$
	$m_B'(a_2 \cup a_4) = 0.5$	$m_B'(\Theta) = 0.5$

The value of K^{-1} is computed from Eq. (5-17) as unity minus the sum of the products of the probability masses assigned by the knowledge sources to members of the empty set. Thus, K^{-1} is given by

$$K^{-1} = 1 - 0.30 = 0.70, \tag{5-19}$$

and its inverse K by

$$K = 1.429. \tag{5-20}$$

As shown in Table 5.6, the probability mass corresponding to the null set element is set to zero and the probability masses of the nonempty set elements are multiplied by K so that their sum is unity. In this example, a type 1 target is declared, but its friendly or hostile nature is undetermined.

Table 5.6: Probability masses of nonempty set elements increased by *K*

$m_A(\Theta) = 0.4$	$m(a_2 \cup a_4) = 0.286$	$m(\Theta) = 0.286$
$m_A(a_1 \cup a_3) = 0.6$	0	$m(a_1 \cup a_3) = 0.429$
	$m_B'(a_2 \cup a_4) = 0.5$	$m_B'(\Theta) = 0.5$

When three or more sensors contribute information, the application of Dempster's rule is repeated using the inner elements calculated from the first application of the rule as the new first column and the probability masses from the next sensor as the entries for the last row (or vice versa).

5.5 Comparison of Dempster-Shafer with Bayesian decision theory

Dempster-Shafer decision theory is more general than Bayesian in the sense that Dempster-Shafer permits probabilities that express certainty or confidence to be assigned directly to an uncertain event, namely, any of the propositions in the frame of discernment Θ. Furthermore, Dempster-Shafer allows sensor classification error to be represented by a probability assignment directly to an uncertainty class Θ. Dempster-Shafer also allows assignment of probability mass to the union of the propositions in the frame of discernment. Bayesian theory permits probabilities to be assigned only to the original propositions themselves. This is expressed mathematically in Bayesian inference as

$$P(a + b) = P(a) + P(b) \tag{5-21}$$

under the assumption that a and b are disjoint propositions. In Dempster-Shafer,

$$P(a + b) = P(a) + P(b) + P(a \cup b). \tag{5-22}$$

Shafer expresses the limitation of Bayesian theory in a more general way: "Bayesian theory cannot distinguish between lack of belief and disbelief. It does not allow one to withhold belief from a proposition without according that belief to the negation of the proposition."

Bayesian theory does not have a convenient representation for ignorance or uncertainty. Prior distributions have to be known or assumed with Bayesian. A Bayesian support function[6] ties all of its probability mass to single points in Θ.

There is no freedom of motion, i.e., no uncertainty interval. The user of a Bayesian support function must somehow divide the support among singleton propositions. This may be easy in some situations such as an experiment with a fair die. If we believe a fair die shows an even number, we can divide the support into three parts, namely, 2, 4, and 6. If the die is not fair, then Bayesian theory does not provide a solution.

Thus, the difficulty with Bayesian theory is in representing what we actually know without being forced to overcommit when we are ignorant. With Dempster-Shafer, we use information from the sensors (knowledge sources) to find the support available for each proposition. For the fair-die example, Dempster-Shafer gives the probability mass m_k(even). If the die were not fair, Dempster-Shafer would still give the appropriate probability mass.

Therefore, there is no inherent difficulty in using Bayesian statistics when the required information is available. However, when knowledge is not complete, i.e., ignorance exists about the prior probabilities associated with the propositions in the frame of discernment, Dempster-Shafer offers an alternative approach. The Dempster-Shafer formulation of a problem collapses into the Bayesian when the uncertainty interval is zero for all propositions and the probability mass assigned to unions of propositions is zero. Waltz and Llinas present an example for the fusion of identification-friend-foe and electronic support measure sensor data to show that the Bayesian approach takes less computation time than Dempster-Shafer to achieve a given belief or probability level. The time difference may or may not be significant, depending on the tactical situation.[7]

Research into generalized evidence processing (GEP) theory, which separates the hypotheses (propositions) from the decisions, allows Bayesian decisions to be extended into a frame of discernment that incorporates multiple hypotheses. With GEP, evidence from nonmutually exclusive propositions can be combined in a Bayesian formulation to reach a decision. The rules in GEP for combining evidence from multiple sensors are analogous to those of Dempster.[8,9,10]

5.6 Summary

The Dempster-Shafer approach to object detection, classification, and identification allows each sensor to contribute information to the extent of its knowledge. Incomplete knowledge about propositions that represent objects in a sensor's field of view is accounted for by assigning a portion of the sensor's probability mass to the uncertainty class. Dempster-Shafer can also assign probability mass to the union of propositions if the evidence supports it. It is in these regards that Dempster-Shafer differs from Bayesian decision theory because Bayesian theory does not have a representation for uncertainty and permits probabilities to be assigned only to the original propositions themselves.

Probability mass may be used to define an uncertainty interval that expresses the support and plausibility for a proposition. Support is the sum of direct sensor evidence for the proposition. Plausibility is the sum of all probability mass not directly assigned by the sensor to the negation of the proposition. Examples were presented to show how probability mass assigned by a sensor to various propositions is used to calculate and interpret the uncertainty interval.

Dempster's rule provides the formalism to combine probability masses from different sensors or information sources. The intersection of propositions with the largest probability mass is selected as the output of the fusion process. If the intersections of the propositions form a null set, the probability masses of the null set elements are redistributed among the nonempty set members.

References

1. R.A. Dillard, *Computing Confidences in Tactical Rule-Based Systems by Using Dempster-Shafer Theory*, NOSC TD 649 (AD A 137274), Naval Ocean Systems Center, San Diego, CA, Sept. 14, 1983.
2. S.S. Blackman, *Multiple Target Tracking with Radar Applications*, Artech House, Norwood, MA (1986).
3. G. Shafer, *A Mathematical Theory of Evidence*, Princeton University Press, Princeton, NJ (1976).
4. P.L. Bogler, "Shafer-Dempster reasoning with applications to multisensor target identification systems," *IEEE Trans. Syst., Man, Cybern.*, SMC-17(6), 968-977 (1987).
5. T.D. Garvey, J.D. Lowrance, and M.A. Fischler, "An inference technique for integrating knowledge from disparate sources," *Proc. Seventh International Joint Conference on Artificial Intelligence*, Vol. I, IJCAI-81, 319-325 (1981).
6. J.A. Barnett, "Computational methods for a mathematical theory of evidence," *Proc. Seventh International Joint Conference on Artificial Intelligence*, Vol. II, IJCAI-81, 868-875 (1981).
7. E. Waltz and J. Llinas, *Multisensor Data Fusion*, Artech House, Norwood, MA (1990).
8. S.C.A. Thomopoulos, R. Viswanathan, and D.C. Bougoulias, " Optimal decision fusion in multiple sensor systems," *IEEE Trans. Aerospace Elect. Systems*, AES-23(5), 644-653 (Sept. 1987).
9. S.C.A. Thomopoulos, "Theories in distributed decision fusion: comparison and generalization," *Sensor Fusion III: 3-D Perception and Recognition, Proc. SPIE* 1383, 623-634 (1990).
10. S.C.A. Thomopoulos, "Sensor Integration and Data Fusion," *J. Robotic Syst.*, 7(3), 337-372 (June 1990).

Chapter 6

Artificial Neural Networks

Biological systems perform pattern recognition using interconnections of large numbers of cells called neurons. The large number of parallel neural connections make the human information processing system adaptable, context-sensitive, error-tolerant, large in memory capacity, and real-time responsive. These characteristics of the human brain provide an alternative model to the more common serial, single-processor signal processing architecture. Although each human neuron is relatively slow in processing information (on the order of milliseconds), the overall processing of information in the human brain is completed in a few hundred milliseconds. The processing speed of the human brain suggests that biological computation involves a small number of serial steps, each massively parallel. Artificial neural networks attempt to mimic the perceptual or cognitive power of humans using the parallel-processing paradigm. Table 6.1 compares the features of artificial neural networks and the more conventional von Neumann serial signal processing architecture.

Table 6.1: Comparison of artificial neural network and von Neumann architectures

Neural Network	von Neumann
No separate arithmetic and memory units and thus no von Neumann bottleneck	Separate arithmetic and memory units
Simple devices densely interconnected	Many microcomputers connected in parallel
Programmed by specifying the architecture and the learning rules used to modify the interconnection weights	Programmed with high-level, assembly, or machine languages
Finds approximate solutions quickly	Must be specifically programmed to find each type of desired solution
Fault tolerance may be achieved through the normal artificial neural network architecture	Fault tolerant through specific programming or use of parallel computers

6.1 Applications of artificial neural networks

Neural network applications include recognition of visual images of shapes and orientations under varied conditions; speech recognition where pitch, rate, and volume vary from sample to sample; and adaptive control. These applications typically involve character recognition, image processing, and direct and parallel implementations of matching and search algorithms.[1,2]

Artificial neural networks can be thought of as using a trainable nonalgorithmic, blackbox approach suitable for solving problems that are generally ill defined and require large amounts of processing through massive parallelism. These problems possess the following characteristics:

- A high-dimensional problem space;
- Complex interactions between problem variables;
- A solution space that may be empty, contain a unique solution, or (most typically) contain a number of useful solutions.

The computational model provided by artificial neural networks has the following attributes:

- A variable interconnection of simple elements or units;
- A learning approach based on modifying interelement connectivity as a function of training data;
- Use of a training process to store information in an internal structure that enables the network to correctly classify new similar patterns and thus exhibit the desired associative or generalization behavior;
- A dynamic system whose state (e.g., unit outputs and interconnection weights) changes with time in response to external inputs or an initial unstable state.

6.2 Adaptive linear combiner

The basic building block of nearly all artificial neural networks is the adaptive linear combiner[1] shown in Figure 6.1. Its output s_k is a linear combination of all its inputs. In a digital implementation, an input signal vector or input pattern vector $\boldsymbol{X}_k = [x_{0k}, x_{1k}, x_{2k}, \ldots, x_{nk}]^{\mathrm{T}}$ and a desired response d_k (a known response to the special input used to train the combiner) are applied at time k. The symbol T indicates a transpose operation. The components of the input vector are weighted by a set of coefficients called the weight vector $\boldsymbol{W}_k = [w_{0k}, w_{1k}, w_{2k}, \ldots, w_{nk}]^{\mathrm{T}}$. The output of the network is given by the weighted input vector, denoted by the inner product $s_k = \boldsymbol{X}_k^{\mathrm{T}} \boldsymbol{W}_k$. The components of $\boldsymbol{X}_k$ may be either analog or binary. The weights are continuously variable positive or negative numbers.

During the training process, a number of input patterns and corresponding desired responses are presented to the linear combiner. An adaptation algorithm is used to automatically adjust the weights so that the output responses to the input patterns are as close as possible to their respective desired responses. The simple least mean square (LMS) algorithm is commonly used to adapt the weights in linear neural networks. This algorithm evaluates and minimizes the sum of squares of the linear errors ε_k over the training pattern set. The linear error is defined as the difference between the desired response d_k and the linear output s_k at time k.

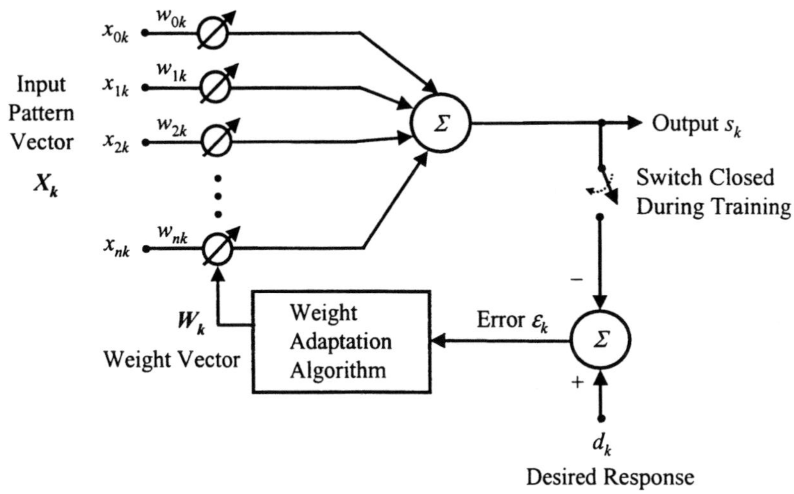

Figure 6.1: Adaptive linear combiner

6.3 Linear classifiers

Both linear and nonlinear artificial neural networks have been developed. The nonlinear classifiers can correctly classify a larger number of input patterns and are not limited to only linear separable forms of pattern discrimination. They are discussed later in the chapter.

One type of linear classifier used in many artificial neural networks is the adaptive linear element or Adaline developed by Widrow and Hoff.[3] This adaptive threshold logic device contains an adaptive linear combiner cascaded with a hard-limiting quantizer as shown in Figure 6.2. Adalines may also be constructed without the nonlinear output device. The quantizer produces a binary ±1 output $y_k = \text{sgn}(s_k)$ where sgn represents the signum function $s_k/|s_k|$. The bias weight w_{0k} connected to the constant input $x_0 = +1$ controls the threshold level of the quantizer.

An adaptive algorithm is used to adjust the weights of the Adaline so that it responds correctly to as many input patterns as possible in a training set that has binary desired responses. Once the weights are adjusted, the response of the trained Adaline is tested by applying new input patterns that were not part of the training set. If the Adaline produces correct responses with some high probability, then generalization is said to have occurred.

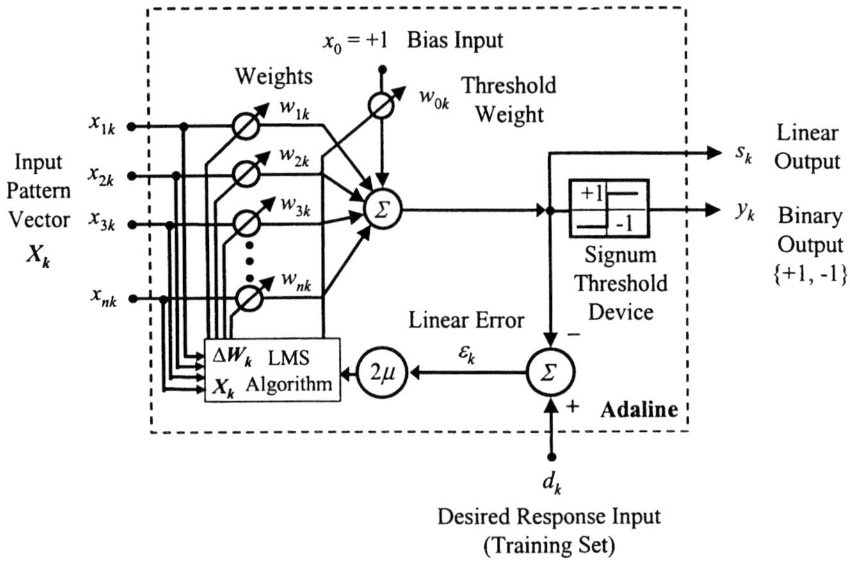

Figure 6.2: Adaptive linear element (Adaline)

6.4 Capacity of linear classifiers

The average number of random patterns with random binary desired responses that an Adaline can learn to classify correctly is approximately equal to twice the number of weights. This number is called the statistical pattern capacity C_s of the Adaline. Furthermore, the probability that a training set is linearly separable is a function of the number N_p of input patterns in the training set and the number N_w of weights including the threshold weight. The probability of linear separability is plotted in Figure 6.3 as a function of the ratio N_p to N_w for several values of N_w. As the number of weights increases, the statistical pattern capacity of the Adaline, $C_s = 2N_w$, becomes an accurate estimate of the number of responses it can learn.[4]

Figure 6.3 also demonstrates that a problem is guaranteed to have a solution if the number of patterns is equal to or less than half of the statistical pattern capacity, i.e., if the number of patterns is equal to or less than the number of weights. This number of patterns is called the deterministic pattern capacity C_d of the Adaline. The capacity results apply to randomly selected training patterns. Since the training set patterns in most problems of practical interest are not random, but exhibit some statistical regularity, the number of patterns learned often far exceeds the statistical capacity. The increase in the number of learned patterns is due to the regularities that make generalization possible, allowing the Adaline to learn many of the training patterns before they are even presented.

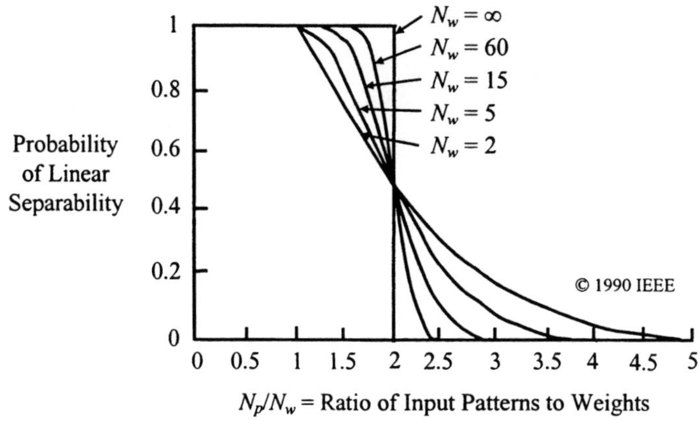

Figure 6.3: Probability of training pattern separation by an Adaline
(B. Widrow and M. A. Lehr, "30 years of adaptive neural networks: perceptron, Madaline, and backpropagation," *Proc. IEEE*, 78(9), 1415-1442 [Sept. 1990])

6.5 Nonlinear classifiers

The nonlinear classifier possesses increased capacity and the ability to separate patterns that have nonlinear boundaries. The two types of nonlinear classifiers described below are the multiple adaptive linear element classifier or Madaline and the multielement feedforward network.

6.5.1 Madaline

The Madaline was originally used to analyze retinal stimuli by connecting the inputs to a layer of adaptive Adalines whose outputs were connected to a fixed logic device that generated the output. An adaptation of this network is illustrated in Figure 6.4 using two Adalines connected to an AND threshold logic output device.

Other types of Madalines may be constructed with many more inputs, many more Adalines in the first layer, and with various logic devices in the second layer. Although the adaptive elements in the original Madalines used the hard-limiting signum quantizers, other nonlinear networks, including the backpropagation network discussed later in this chapter, use differentiable nonlinearities such as sigmoid or S-shaped functions illustrated in Figure 6.5.

The input-output relation for the signum function is denoted by

$$y_k = \text{sgn}(s_k), \tag{6-1}$$

where

$$\text{sgn}(s_k) = \frac{s_k}{|s_k|} \tag{6-2}$$

and s_k and y_k are the linear and binary outputs of the network, respectively.

For the sigmoid function, the input-output relation is given by

$$y_k = \text{sgm}(s_k). \tag{6-3}$$

A typical sigmoid function is represented by the hyperbolic tangent as

$$y_k = \tanh(s_k) = (1 - e^{-2s_k})/(1 + e^{-2s_k}). \tag{6-4}$$

However, sigmoid functions can be generalized in neural network applications to include any smooth nonlinear function at the output of a linear adaptive element.[2]

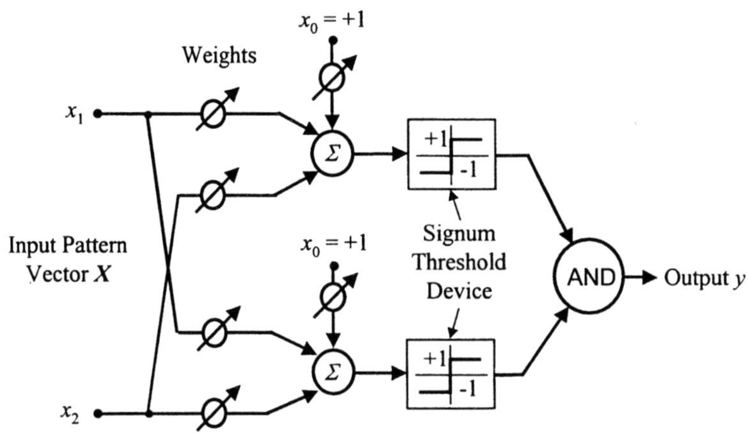

Figure 6.4: Madaline constructed of two Adalines with an AND threshold logic output

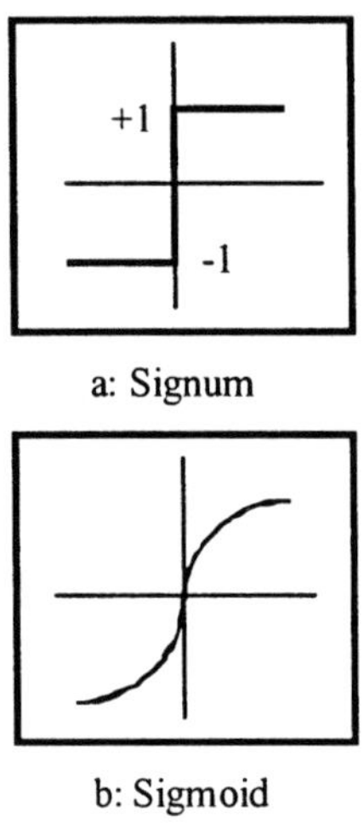

Figure 6.5: Threshold functions used in artificial neural networks

Figure 6.6 shows implementations of three threshold logic output functions, namely, AND, OR, and MAJORITY vote taker. The weight values in the figure implement these three functions, but the weights are not unique.

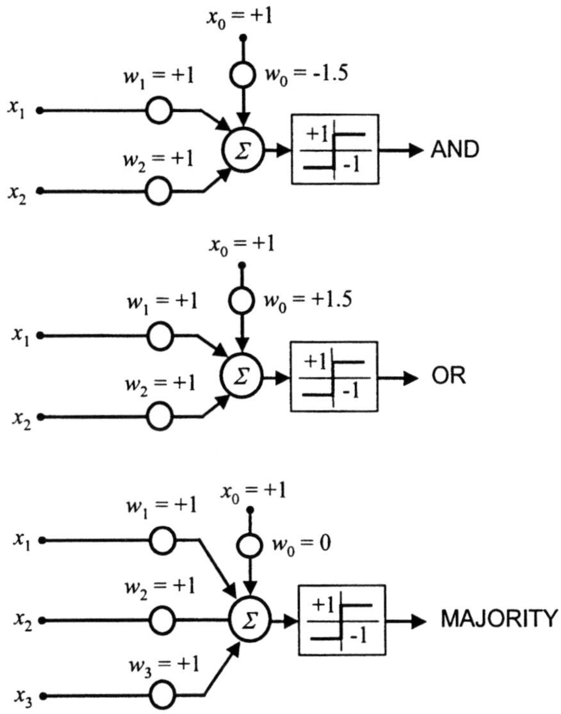

Figure 6.6: Fixed-weight Adaline implementations of AND, OR, and MAJORITY threshold logic functions

6.5.2 Feedforward network

Typical feedforward networks have many layers and usually all are adaptive. A fully connected, three-layer feedforward network is illustrated in Figure 6.7. Adalines are used in this illustration to represent an artificial neuron-processing element that connects inputs to a summing node. The inputs are subject to modification by the adjustable weights. The output of the summing node may then pass through a hard or soft limiter. In a fully connected network, each processing element receives inputs from every output in the preceding layer. During training, the response of each output element in the network is compared with a corresponding desired response. Error signals associated with the output elements are easily computed, allowing for straightforward adaptation or training of the output layer. However, obtaining error signals for hidden-layer processing elements, i.e., elements in layers other than the output layer, requires more complex learning rules such as the backpropagation algorithm.

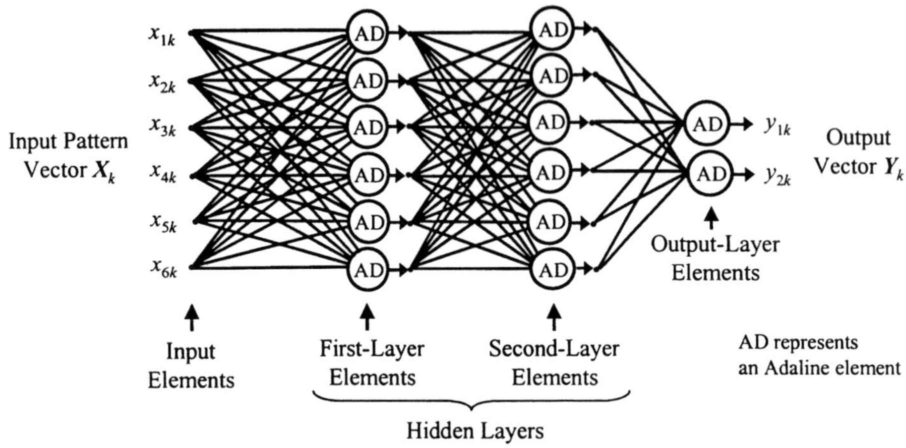

Figure 6.7: Three layer fully connected feedforward neural network

In general, a feedforward network is composed of a hierarchy of processing elements. The processing elements are organized in a series of two or more mutually exclusive sets of layers. The input elements are a holding place for the values applied to the network. These elements do not implement a separate mapping or conversion of input data and their weights are insignificant. The last, or output layer, permits the final state of the network to be read. Between these two extremes are zero or more layers of hidden elements. The hidden layers remap the inputs and results of previous layers and, thereby, produce a more separable or more easily classifiable representation of the data. In the architecture of Figure 6.7, links or weights connect each element in one layer to only those in the next higher layer. An implied directionality exists in these connections, whereby the output of one element, scaled by the connecting weight, is fed forward to provide a portion of the activation for the elements in the next higher layer. Forms of feedforward networks, other than that of Figure 6.7, have been developed. In one, the processing elements receive signals directly from each input component and from the output of each preceding processing element.[1]

6.6 Capacity of nonlinear classifiers

The average number of random patterns, having random binary responses, that a Madeline network represented by Figure 6.4 can learn to classify is equal to the capacity per Adaline, or processing element, multiplied by the number of Adalines in the network. Therefore, the statistical capacity C_s of the Madaline is approximately equal to twice the number of adaptive weights. Although the Madeline and the Adeline have roughly the same capacity per adaptive weight, the Madaline can separate sets with nonlinear separation boundaries.

The capacity of a feedforward signum network with an arbitrary number of layers is dependent[5] on the number of weights N_w and the number of outputs N_y. For a two-layer fully connected feedforward network of signum Adalines with N_x inputs (excluding bias inputs) and N_y outputs, the minimum number of weights N_w is bounded by

$$\frac{N_y N_p}{1+\log_2 N_p} \leq N_w < N_y \left(\frac{N_p}{N_x}+1\right)(N_x+N_y+1)+N_y \tag{6-5}$$

when the network is required to learn to map any set of N_p patterns in the general position[1] into any set of binary desired response vectors with N_y outputs. The statistical and deterministic capacities given above for the linear classifier are also dependent upon the input patterns being in general position. If the patterns are not in general position, the capacity results represent upper bounds to the actual capacity that can be obtained.[1,4]

For a two-layer feedforward signum network with at least five times as many inputs and hidden elements as outputs, the deterministic pattern capacity is bounded *below* by a number slightly smaller than N_w/N_y. For any feedforward network with a large ratio of weights to outputs (at least several thousand), the deterministic pattern capacity is bounded *above* by a number slightly larger than $N_w/N_y \log_2(N_w/N_y)$. Thus, the deterministic pattern capacity C_d of a two-layer network is bounded by

$$(N_w/N_y) - K_1 \leq C_d \leq N_w/N_y \log_2(N_w/N_y) + K_2, \tag{6-6}$$

where K_1 and K_2 are positive numbers that are small if the network is large with few outputs relative to the number of inputs and hidden elements. Equation (6-6) also bounds the statistical capacity of a two-layer signum network.

The following rules of thumb are useful for estimating pattern capacity:

- Single-layer network capacities are used as capacity estimates for multilayer networks;
- The capacity of sigmoid (soft limiting) networks cannot be less than that of signum networks of equal size;

[1]. Patterns are in general position with respect to an Adaline that does not contain a threshold weight if any subset of pattern vectors that contains no more than N_w members forms a linearly independent set. Equivalently, the patterns are in general position if no set of N_w or more input points in the N_w-dimensional pattern space lie on a homogeneous hyperplane. For an Adaline with a threshold weight, general position occurs when no set of N_w or more patterns in the $(N_w - 1)$-dimension pattern space lie on a hyperplane not constrained to pass through the origin.

- For good generalization, i.e., classification of patterns not presented during training, the training set pattern size should be several times larger than the network's capacity such that $N_p >> N_w/N_y$.

Finding the optimum number of hidden elements for a feedforward network is problem dependent and often involves considerable engineering judgment. While intuition may suggest that more hidden elements will improve the generalization capability of the network, excessively large numbers of hidden elements may be counterproductive. A difficulty that may arise if the number of hidden elements grows too large is depicted in Figure 6.8.

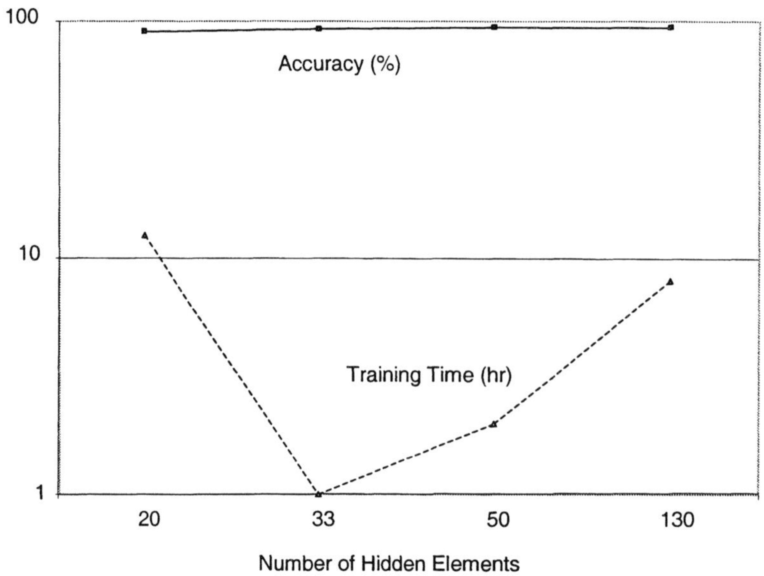

Figure 6.8: Effect of number of hidden elements on feedforward neural network training time and output accuracy for a specific problem
(Adapted from R. Gaborski, "An intelligent character recognition system based on neural networks," *Research Magazine*, Eastman Kodak Company, Rochester, NY [Spring 1990] Reprinted with permission from Eastman Kodak Company.)

Here the accuracy of the output decision rapidly approaches a limiting value and the training time rapidly falls when the number of hidden elements is kept below some value that is problem specific. As the number of hidden elements is increased further, the training time increases rapidly, while the accuracy grows much more slowly.[2] The specific values shown in this figure are not general results, but rather apply to a specific neural network application.[6]

6.7 Supervised and unsupervised learning

The description of learning algorithms as supervised or unsupervised originates from pattern recognition theory. Supervised learning uses pattern class information; unsupervised learning does not. Learning seeks to accurately estimate $p(X)$, the probability density function that describes the continuous distribution of patterns X in the pattern space. The supervision in supervised learning provides information about $p(X)$. However, the information may be inaccurate. Unsupervised learning makes no assumptions about $p(X)$. It uses minimal information.[7]

Supervised learning algorithms depend on the class membership of each training sample x. Class membership information allows supervised learning algorithms to detect pattern misclassifications and compute an error signal or vector, which reinforces the learning process.

Unsupervised learning algorithms use unlabeled pattern samples and blindly process them. Unsupervised learning algorithms often have less computational complexity and less accuracy than supervised learning algorithms.[7] Unsupervised learning algorithms learn quickly, often on a single pass of noisy data. Thus, unsupervised learning is applied to many high-speed real-time problems where time, information, or computational precision is limited.

Examples of supervised learning algorithms include the steepest descent and error correction algorithms that estimate the gradient or error of an unknown mean-squared performance measure. The error depends on the unknown probability density function $p(X)$.

Unsupervised learning may occur in several ways. It may adaptively form clusters of patterns or decision classes that are defined by their centroids. Other unsupervised neural networks evolve attractor basins in the pattern state space. Attractor basins correspond to pattern classes and are defined by their width, position, and number.

6.8 Supervised learning rules

Figure 6.9 shows the taxonomy used by Widrow and Lehr to summarize the supervised learning rules developed to train artificial neural networks that incorporate adaptive linear elements.[1] The rules are first separated into steepest descent and error correction categories, then into layered network and single element categories, and finally into nonlinear and linear rules.

Steepest descent or gradient rules alter the weights of a network during each pattern presentation with the objective of reducing mean square error (MSE) averaged over all training patterns. Although other gradient approaches are available, MSE remains the most popular. Error correction rules, on the other hand, alter the weights of a network to reduce the error in the output response to the current training pattern. Both types of rules use similar training procedures. However, since they are based on different objectives, they may have significantly different learning characteristics. Error correction rules are most often applied when training objectives are not easily quantified or when a problem does not lend itself to tractable analysis.

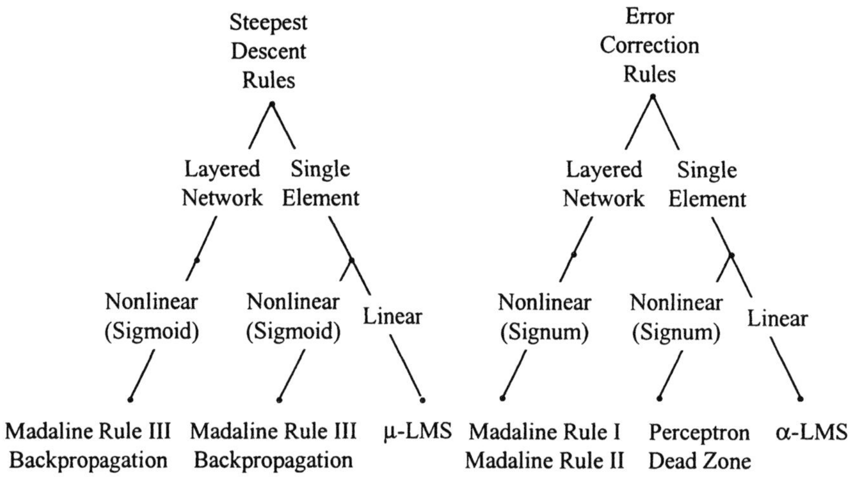

Figure 6.9: Learning rules for artificial neural networks that incorporate adaptive linear elements

(Adapted from B. Widrow and M. A. Lehr, "30 years of adaptive neural networks: perceptron, Madaline, and backpropagation," *Proc. IEEE*, 78(9), 1415-1442 [Sept. 1990])

6.8.1 μ-LMS steepest descent algorithm

The μ-LMS steepest descent algorithm performs approximate steepest descent on the MSE surface in weight space. Since this surface is a quadratic function of the weights, it is convex in shape and possesses a unique minimum. Steepest descent algorithms adjust the network weights by computing or estimating the error between the network output and the desired response to a known input. The weight adjustment is proportional to the gradient formed by the partial derivative of the error with respect to the weight, but in the direction opposite to the gradient.

The algebraic expression for updating the weight vector is given by

$$\boldsymbol{W}_{k+1} = \boldsymbol{W}_k + 2\mu\, \varepsilon_k \boldsymbol{X}_k. \tag{6-7}$$

Stability and speed of convergence are controlled by the learning constant μ. If μ is too small, the μ-LMS algorithm moves very slowly down the estimated mean square error surface and learning may be prohibitively slow. If μ is too large, the algorithm may leap recklessly down the estimated mean square error surface and the learning may never converge. In this case, the weight vector may land randomly at points that correspond to first larger and then smaller values of the total mean square error surface.[7]

The learning constant should vary inversely with system uncertainty. The more uncertain the sampling or training environment, the smaller should be the value of μ to avoid divergence of the training process. The learning constant can be larger to speed convergence when there is less uncertainty in the sampling environment.

If the input patterns are independent over time, the mean and variance of the weight vector converge for most practical purposes if

$$0 < \mu < 1/\text{trace}\,[\boldsymbol{R}_k], \tag{6-8}$$

where trace $[\boldsymbol{R}_k]$ equals the sum of the diagonal elements of $\boldsymbol{R}_k$ which, in turn, is equal to the average signal power of the $\boldsymbol{X}_k$-vector or $E[\boldsymbol{X}_k^T \boldsymbol{X}_k]$. The variable $\boldsymbol{R}_k$ may also be viewed as the autocorrelation matrix of the input vectors $\boldsymbol{X}_k$ when the input patterns are independent.

6.8.2 α-LMS error correction algorithm

Using a fixed input pattern, the α-LMS algorithm optimizes the weights to reduce the error between the network output and the desired response by a factor α. The weight vector update is found as

$$\boldsymbol{W}_{k+1} = \boldsymbol{W}_k + \alpha\, \varepsilon_k \frac{\boldsymbol{X}_k}{|\boldsymbol{X}_k|^2} \tag{6-9}$$

and the error reduction factor as

$$\Delta\varepsilon_k = -\alpha\, \varepsilon_k. \tag{6-10}$$

The negative sign indicates that the change in error is in the direction opposite to the error itself. Stability and speed of convergence of the algorithm are controlled by the value of α. When the input pattern vectors are independent over time, stability is ensured for most practical purposes when $0 < \alpha < 2$. Values of α greater than 1 overcorrect the error, while total error correction corresponds to $\alpha = 1$. A practical range for α lies between 0.1 and 1.0. When all input patterns are equal in length, the α-LMS algorithm minimizes mean square error and is best known for this property.

6.8.3 Comparison of the μ-LMS and α-LMS algorithms

Both the μ-LMS and α-LMS algorithms rely on the least mean square instantaneous gradient for their implementation. The α-LMS is self-normalizing, with α determining the fraction of the instantaneous error to be corrected with each iteration, whereas μ-LMS is a constant coefficient linear algorithm that is easier to analyze. The α-LMS is similar to the μ-LMS with a continually variable learning constant. Although the α-LMS is somewhat more difficult to implement and analyze, experiments show that it is a better algorithm than the μ-LMS when the eigenvalues of the input autocorrelation function matrix $\boldsymbol{R}$ are highly disparate. In this case, the α-LMS gives faster convergence for a given difference between the gradient estimate and the true gradient. This difference is propagated into the weights as "gradient noise." The μ-LMS has the advantage that it will always converge in the mean to the minimum MSE solution, while the α-LMS may converge to a somewhat biased solution.[1]

6.8.4 Madaline I and II error correction rules

The Madaline I error correction training rule applies to a two-layer Madaline network such as the one depicted in Figure 6.4. The first layer consists of hard-limited signum Adaline elements. The outputs of these elements are connected to a second layer containing a single fixed-threshold logic element, e.g., AND, OR, or MAJORITY vote taker. The weights of the Adalines are initially set to small random values. The Madaline I rule adapts the input elements in the first layer such that the output of the threshold logic element is in the desired state as specified by a training pattern. No more Adaline elements are adapted than necessary to correct the output decision. The elements whose linear outputs are nearest to zero are adapted as they require the smallest weight changes to reverse their output responses. Whenever an Adaline is adapted, the weights are changed in the direction of its input vector because this provides the required error correction with minimal weight change.

The Madaline II error correction rule applies to multilayer binary networks with signum thresholds. Training is similar to training with the Madaline I algorithm. The weights are initially set to small random values. Training patterns are presented in a random sequence. If the network produces an error during training, the first-layer Adaline with the smallest linear output is adapted first by inverting its binary output. If the number of output errors produced by the training patterns is reduced by the trial adaptation, the weights of the selected elements are changed by the α-LMS error correction algorithm in a direction that reinforces the bit reversal with minimum disturbance to the weights. If the trial adaptation does not improve the network response, the weight adaptation is not performed. After finishing with the first element, other Adalines in the first layer with sufficiently small linear outputs are adapted. After exhausting all possibilities in the first layer, the next layer elements are adapted, and so on. When the final layer is reached and all appropriate elements have been adapted by the α-LMS algorithm, a new training pattern is selected at random and the procedure is repeated.

6.8.5 Perceptron rule

In some cases, the α-LMS algorithm may fail to separate training patterns that are linearly separable. In these situations, nonlinear rules such as Rosenblatt's α-perceptron rule may be suitable.[8]

The perceptron, shown in Figure 6.10, is a feedforward network with one output neuron that learns a separating hyperplane in pattern space. The first layer of fixed threshold logic devices processes a number of input patterns that are sparsely and randomly connected to it. The outputs of the first layer feed a second layer composed of a single adaptive linear threshold element or neuron. The adaptive element is similar to the Adaline, with two exceptions: its input signals are binary {0, 1} and no bias weight is used.

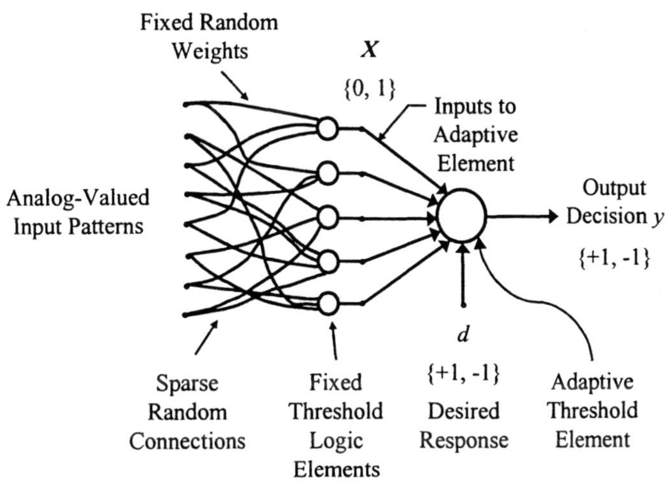

Figure 6.10: Rosenblatt's perceptron

The adaptive threshold element of the perceptron is illustrated in Figure 6.11. If the output decision y_k disagrees with the desired binary response d_k, adaptation occurs by adding the input vector to the weight vector of the adaptive threshold element when the quantizer error is positive and subtracting the input vector from the weight vector when the error is negative. The quantizer error shown in Figure 6.11 by $\tilde{\varepsilon}_k$ is given by

$$\tilde{\varepsilon}_k = d_k - y_k. \tag{6-11}$$

The perceptron rule is identical to the α-LMS algorithm except that the perceptron uses half the quantizer error, $\tilde{\varepsilon}_k/2$, in place of the normalized linear error $\varepsilon_k/|X_k|^2$ of the α-LMS algorithm. Thus, the perceptron rule gives the weight vector update as

$$W_{k+1} = W_k + \alpha(\tilde{\varepsilon}_k/2)X_k. \tag{6-12}$$

Normally α is set equal to 1. In contrast to α-LMS, α does not affect the stability of the perceptron algorithm. It affects the convergence time only if the initial weight vector is nonzero. While the α-LMS algorithm may be applied to either analog or binary desired responses, the perceptron may only be used with binary desired responses. Although the perceptron was developed in the late 1950s, its widespread application was not extensive because its classification ability was dependent on training with linearly separable patterns and a training algorithm for the multilayer case did not exist. The multilayer feedforward networks and the backpropagation algorithm have helped to remedy these constraints.

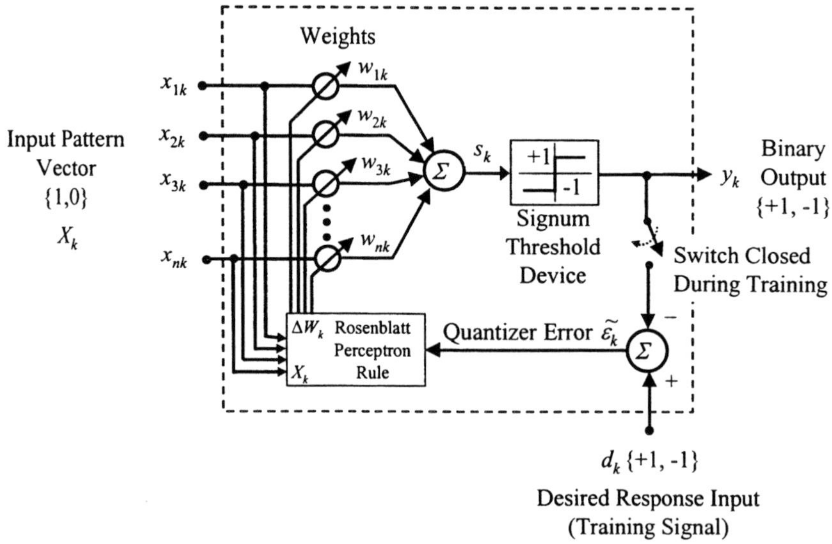

Figure 6.11: Adaptive threshold element of perceptron

6.8.6 Backpropagation algorithm

The backpropagation algorithm is a stochastic steepest descent learning rule used to train single- or multiple-layer nonlinear networks. The algorithm overcomes some limitations of the perceptron rule by providing a framework for computing the weights of hidden layer neurons. The algorithm's stochastic nature implies a search for a random minimum square error separating surface rather than an unknown deterministic mean square error surface. Therefore, the back-propagation algorithm may converge to a local error minima or may not converge at all if a poor choice of initial weights is made. The backpropagation algorithm reduces to the μ-LMS algorithm if all neural elements are linear and if the feedforward topology from input to output layers contains no hidden neurons.

Expressed in biological nomenclature, the backpropagation algorithm recursively modifies the synapses between neuronal fields, i.e., input, hidden, and output layers. The algorithm first modifies the synapses between the output layer and the penultimate layer of hidden or interior neurons. Then the algorithm applies this information to modify the synapses between the penultimate hidden layer and the preceding hidden layer, and so on, until the synapse between the first hidden layer and the input layer is reached.

Training begins by presenting an input pattern vector $\boldsymbol{X}$ to the network. The input values in $\boldsymbol{X}$ are swept forward through the network to generate an output response vector $\boldsymbol{Y}$ and to compute the errors at the output of each layer, including the hidden layers. The effects of the errors are then swept backwards through the network. The backward sweep (1) associates a square error derivative δ with each network element in each layer, (2) computes a gradient from each δ, and (3) updates the weights of each element based upon the gradient for that layer and element. A new pattern is then presented to the network and the process is repeated. Training continues until all patterns in the training set are exhausted. The objective of the backpropagation algorithm is not to reduce the square error derivatives at each layer in the network. Rather the goal is to reduce the sum squared error (the sum of the squares of the difference between the desired response and actual output at each element in the output layer) at the network output.

In applying the backpropagation algorithm, the initial weights are normally set to small random numbers. Multilayer networks are sensitive to the initial weight selection, and the algorithm will not function properly if the initial weight values are either zero or poorly chosen nonzero values. In fact, the network may not learn the set of training examples. If this occurs, learning should be restarted with other values for the initial random weights. The speed of training is affected by the learning constant that appears in the weight update equations. When broad minima that yield small gradients are present, a larger value of the learning constant gives more rapid convergence. For applications with steep and narrow minima, a small value of the learning constant avoids overshooting the solution. Thus, the learning constant should be chosen experimentally to suit each problem. Learning constants between 10^{-3} and 10 have been reported in the literature.[9] Large learning constants can dramatically increase the learning speed, but the solution may overshoot and not stabilize at any network minimum error.

Calculations associated with the backward sweep through the network are roughly as complex as those associated with the forward pass. The algorithm uses different expressions for updating the output layer weights and the hidden layer weights.[1,2,7]

6.8.7 Madaline III steepest descent rule

The Madaline III steepest descent rule is used in networks containing sigmoid Adalines. It avoids some of the problems that occur when backpropagation is used with inaccurate realizations of sigmoid functions and their derivatives. Madaline Rule III works similarly to Madaline Rule II. All the Adalines in the Madaline Rule III network are adapted, but those whose analog sums are closest to zero will usually be adapted most strongly since the sigmoid has its maximum slope at zero, which contributes to high gradient values. As with Madaline Rule II, the objective is to change the weights for the given input training pattern to reduce the sum square error at the network output. The weight vectors of the Adaline elements are selected for adaptation in the LMS direction according to their capabilities for reducing the sum square error at the output. The weight vectors are adjusted in proportion to a small perturbation signal Δs that is added to the sum s_k at the output of the weight vector (as in Figure 6.11). The effect of the perturbation on output y_k and error $\tilde{\varepsilon}_k$ is noted.

The instantaneous gradient is computed in one of two ways, leading to two forms of the Madaline III algorithm for updating the weights. These are

$$W_{k+1} = W_k - \mu \left[\frac{\Delta(\tilde{\varepsilon}_k)^2}{\Delta s}\right] X_k \tag{6-13}$$

and

$$W_{k+1} = W_k - 2\mu\, \tilde{\varepsilon}_k \left[\frac{\Delta\tilde{\varepsilon}_k}{\Delta s}\right] X_k. \tag{6-14}$$

The learning constant μ is similar to that used in the μ-LMS algorithm. The two forms for updating the weights are equivalent for small perturbations Δs.

6.8.8 Dead zone algorithms

Two algorithms that incorporate dead zones into the training process were developed by Mays.[10] These are the increment adaptation rule and the modified relaxation adaptation rule. Increment adaptation associates a dead zone with the linear outputs s_k, where the dead zone is set equal to $\pm\gamma$ about zero. The dead zone reduces sensitivity to weight errors. If the linear output is outside the dead zone, the weight update follows a variant of the perceptron rule given by

$$W_{k+1} = W_k + \alpha\, \tilde{\varepsilon}_k \frac{X_k}{2|X_k|^2} \quad \text{if } |s_k| \geq \gamma. \tag{6-15}$$

If the linear output is inside the dead zone, the weights are adapted by another variant of the perceptron rule as

$$W_{k+1} = W_k + \alpha\, d_k \frac{X_k}{|X_k|^2} \text{ if } |s_k| < \gamma, \tag{6-16}$$

where $\tilde{\varepsilon}_k$ is given by Eq. (6-11) and d_k is the desired response defined by the training pattern.

Mays proved that if the training patterns are linearly separable, increment adaptation would always converge and separate the patterns in a finite number of steps. If the training set is not linearly separable, the increment adaptation rule typically performs better than the perceptron rule because a sufficiently large dead zone causes the weight vector to adapt away from zero when any reasonably good solution exists.[1]

The modified relaxation adaptation rule uses the linear error ε_k, depicted in Figures 6.1 and 6.2 for the α-LMS algorithm, to update the weights. The modified relaxation rule differs from the α-LMS in that a dead zone is created. If the quantizer output y_k is correct and the linear output s_k falls outside the dead zone, the weights are not updated. In this case

$$W_{k+1} = W_k \text{ if } \varepsilon_k = 0 \text{ and } |s_k| \geq \gamma. \tag{6-17}$$

If the quantizer output is incorrect or if the linear output falls within the dead zone $\pm\gamma$, the weights are updated following the α-LMS algorithm according to

$$W_{k+1} = W_k + \alpha\, \varepsilon_k \frac{X_k}{|X_k|^2} \text{ otherwise.} \tag{6-18}$$

6.9 Other artificial neural networks

Other types of artificial neural networks have been developed in addition to the adaptive linear element (Adaline), multiple adaptive linear element (Madaline), perceptron, and multilayer adaptive linear element feedforward networks. These include the multilayer perceptron, Kohonen self-organizing network, Grossberg adaptive resonance network, counterpropagation network, and Hopfield network. The Kohonen, Grossberg, and counterpropagation networks use unsupervised learning. The characteristics of these networks are summarized[2,11,12,13] in Table 6.2.

6.10 Summary

Artificial neural networks are commonly applied to solve problems that involve complex interactions between input variables. These applications include target classification, speech synthesis, speech recognition, pattern mapping and recognition, data compression, data association, optical character recognition, and system optimization. The adaptive linear combiner is a basic building block of linear and nonlinear artificial neural networks. Nonlinear classifiers generally can correctly classify a larger number of input patterns than linear classifiers. The statistical capacity or number of random patterns that a linear classifier can learn to classify is approximately equal to twice the number of weights in the processing element. The statistical capacity of nonlinear Madaline networks is also equal to twice the number of weights in the processing elements. However, the Madaline contains more than one processing element and, hence, has a greater capacity than the linear classifier. The capacities of more complex nonlinear classifiers, such as multilayer feedforward networks, can be bounded and approximated by the rules discussed in this chapter. Learning or training rules for single element and multilayer linear and nonlinear classifier networks have been described. In supervised training, the network weights are adjusted to minimize the error between the network output and the desired response to a known input. Linear classifier training rules include the μ-LMS and α-LMS algorithms. Nonlinear classifier rules include the perceptron, backpropagation, Madaline, and dead zone algorithms. The backpropagation algorithm permits optimization of not only the weights in output layer elements of feedforward networks, but also those in the hidden layer elements. The training and applications of the multilayer perceptron, Kohonen self-organizing network, Grossberg adaptive resonance network, counterpropagation network, and Hopfield network have been presented. The Kohonen, Grossberg, and counterpropagation networks are examples of systems that use unsupervised learning based on processing unlabeled samples. These systems adaptively cluster patterns into decision classes.

Table 6.2: Properties of other artificial neural networks

Type	Key Operating Principles	Applications
• Multilayer perceptron	• A multilayer feedforward network. • Trained with supervised learning. • Errors are minimized with backward error propagation that updates the weights applied to the input data by the hidden layer and output neuron layers. • Number of weights is equal to the number of hidden-layer neurons.	• Accommodates complex decision regions in the feature space • Target classification • Speech synthesis
• Kohonen network[14,15,16]	• Feedforward network that works with an unsupervised learning paradigm (processes unlabeled data, i.e., data where the desired classification is unknown). • Uses a mathematical transformation to convert input data vectors into output graphs, maps, or clustering diagrams. • Individual neural network clusters self-organize to reflect input pattern similarity. • Overall structure of the network can be viewed as an array of matched filters that competitively adjust input element weights based on current weights and goodness of match of the output to the training set input. • Trained with winner-take-all algorithm. The winning node is rewarded with a weight adjustment, while the weights of the other nodes are unaffected. Winning node is the one whose output cluster most closely matches the input. • Network can also be trained with multiple winner unsupervised learning where the *K* neurons best matching the input vector are allowed to have their weights adjusted. The outputs of the winning neurons can be set to sum to unity.	• Speech recognition

(Table continued on next page)

Table 6.2: Properties of other artificial neural networks (continued)

Type	Key Operating Principles	Applications
• Grossberg adaptive resonance network[17,18]	• Unsupervised learning paradigm that employs feedforward and feedback computations. • Teaches itself new categories and continues storing information without rejecting pieces of information that are temporarily useless as they may be needed later. Pattern or feature information is stored in clusters. • Uses two layers – an input layer and an output layer. The output layer itself has two sublayers: a comparison layer for short-term memory and a recognition layer for long-term memory. • One adaptive resonance theory network learning algorithm (ART1) performs an offline search through encoded clusters, trying to find a sufficiently close match of the input pattern to a stored cluster. If no match is found, a new class is created.	• Pattern recognition • Target classification
• Counter-propagation network[19,20,21]	• The counterpropagation network consists of two layers that map input data vectors into bipolar binary responses (-1, +1). It allows propagation from the input to a classified output, as well as propagation in the reverse direction. • First layer is a Kohonen layer trained in unsupervised winner-take-all mode. Input vectors belonging to the same cluster activate the same neuron in the Kohonen layer. The outputs of the Kohonen layer neurons are binary unipolar values 0 and 1. The first layer organizes data, allowing, for example, faster training to perform associative mapping than is typical of other two-layer networks. • Second layer is a Grossberg layer that orders the mapping of the input vectors into the bipolar binary outputs. The result is a network that behaves as a lookup memory table.	• Target classification • Pattern mapping and association • Data compression

(Table continued on next page)

Table 6.2: Properties of other artificial neural networks (continued)

Type	Key Operating Principles	Applications
• Hopfield network[22,23,24] (a type of associative memory network)	• Associative memories belong to a class of neural networks that learn according to a specific recording algorithm. They usually require *a priori* information and their connectivity (weight) matrices are frequently formed in advance. The network is trained with supervised learning. • Writing into memory produces changes in neural interconnections. Transformation of the input signals by the network allows information to be stored in memory for later output. • No usable addressing scheme exists in associative memory since all memory information is spatially distributed and superimposed throughout the network. • Every neuron is connected to all other neurons. • Network convergence is relatively insensitive to the fraction of elements (15 to 100%) updated at each step. • Each node receives inputs that are processed through a hard limiter. The outputs of the nodes (±1) are multiplied by the weight assigned to the nodes connected by the weight. • The minimum number of nodes is seven times the number of memories to be stored. • The asymptotic capacity C_a of auto-associative networks is bounded by $n/(4 \ln n) < C_a < n/(2 \ln n)$.	• Data association • Optimization problems • Optical character recognition

References

1. B. Widrow and M.A. Lehr, "30 years of adaptive neural networks: perceptron, Madaline, and backpropagation," *Proc. IEEE*, 78(9), 1415-1442 (Sept. 1990).
2. R. Schalkoff, *Pattern Recognition: Statistical, Structural, and Neural Approaches*, John Wiley and Sons, New York, NY (1992).
3. B. Widrow and M.E. Hoff, "Adaptive switching circuits," 1960 IRE Wescon Conv. Rec., Part 4, 96-104 (Aug. 1960). [Reprinted in J.A. Anderson and E. Rosenfeld (Eds.), *Neuro-Computing: Foundations of Research*, MIT Press, Cambridge, MA (1988)].
4. N.J. Nilsson, *Learning Machines*, McGraw-Hill, New York, NY (1965).
5. E.B. Baum, "On the capabilities of multilayer perceptrons," *J. Complex.*, 4, 193-215 (Sept. 1988).
6. R. Gaborski, "An intelligent character recognition system based on neural networks," *Research Magazine*, Eastman Kodak Company, Rochester, NY (Spring 1990).
7. B. Kosko, *Neural Networks and Fuzzy Systems: A Dynamical Systems Approach to Machine Intelligence*, Prentice-Hall, Englewood Cliffs, NJ (1992).
8. F. Rosenblatt, *Principles of Neurodynamics: Pperceptrons and the Theory of Brain Mechanisms*, Spartan Books, Washington, DC (1962).
9. J.M. Zurada, *Introduction to Artificial Neural Systems*, West Publishing Comp., St. Paul, MN (1992).
10. C.H. Mays, *Adaptive threshold logic*, Ph.D. thesis, Tech. Rep. 1557-1, Stanford Electron. Labs., Stanford, CA (Apr. 1963).
11. R.P. Lippmann, "An introduction to computing with neural nets," *IEEE ASSP Mag.*, 4, 4-22 (Apr. 1987).
12. J.A. Anderson and E.Rosenfeld, Eds., *Neurocomputing: Foundations of Research*, M.I.T. Press, Cambridge, MA (1988).
13. B. Kosko, "Unsupervised learning in noise," *IEEE Trans. Neural Networks*, 1(1), 44-57 (March 1990).
14. T. Kohonen, "Self-organized formation of topologically correct feature maps," *Biol. Cybern.*, 43, 59-69 (1982).
15. T. Kohonen, "Analysis of a simple self-organizing process," *Biol. Cybern.*, 44, 135-140 (1982).
16. T. Kohonen, *Self-Organization and Associative Memory*, Springer-Verlag, Berlin (1984).
17. G.A. Carpenter and S. Grossberg, "A massively parallel architecture for a self-organizing neural pattern recognition machine," *Comput. Vis., Graph., Image Process.*, 37, 54-115 (1987).
18. G.A. Carpenter and S. Grossberg, "ART 2: Self-organization of stable category recognition codes for analog input patterns," *Appl. Opt.*, 26(3), 4919-4930 (Dec. 1987).
19. R. Hecht-Nielsen, "Counterpropagation networks," *Appl. Opt.*, 26(3), 4979-4984 (Dec. 1987).

20. R. Hecht-Nielsen, "Applications of counterpropagation networks," *Neural Net.*, 1, 131-139 (1988).
21. R. Hecht-Nielsen, *Neurocomputing*, Addison-Wesley, Reading, MA, 1990.
22. J.J. Hopfield, "Neural networks and physical systems with emergent collective computational abilities," *Proc. Nat. Acad. Sci.*, 79 (Biophysics), 2554-2558 (Apr. 1982).
23. J.J. Hopfield, "Neurons with graded response have collective computational properties like those of two-state neurons," *Proc. Nat. Acad. Sci.*, 81 (Biophysics), 3088-3092 (May 1984).
24. J.J. Hopfield and D.W. Tank, "Computing with neural circuits: a model," *Science*, 233, 625-633 (Aug. 1986).

Chapter 7

Voting Fusion

The fusion configurations shown in Figure 7.1 show the strengths and weaknesses of combining sensor outputs in parallel, series, and in series/parallel. Generally, the parallel configuration provides good detection of targets with suppressed signatures because only one sensor in the suite is required to detect the target. The series configuration provides good rejection of false targets when the sensors respond to signals generated by different phenomena. The weaknesses of these configurations become apparent by reversing their advantages. The parallel is subject to false target detection and susceptibility to decoys, since one sensor may respond to a strong signal from a nontarget. The series arrangement requires signatures to be generated by all the phenomena encompassed by the sensors. Thus, the series configuration functions poorly when one or more of the expected signature phenomena is absent or weak, such as when a target signature is suppressed.

The series/parallel configuration supports a voting fusion process that incorporates the advantages of the parallel and series configurations. These are rejection of signatures from decoys, clutter, and other nontargets and detection of targets that have one or more of their signature domains suppressed. We will show that voting fusion (one of the feature-based inference fusion techniques for object classification) allows the sensors to automatically detect and classify objects to the extent of their knowledge. This process does not require explicit switching of sensors based on the quality of their inputs to the fusion processor or the real-time characteristics of the operating environment. The sensor outputs are always connected to the fusion logic that is designed to incorporate all anticipated combinations of sensor knowledge. Auxiliary operating modes may be added to the automatic voting process to further optimize sensor system performance under some special conditions that are identified in advance. The special conditions may include countermeasures, inclement weather, or high-clutter backgrounds, although the automatic voting may prove adequate in these circumstances as well. Testing and simulation of system performance are needed to ascertain whether auxiliary modes are needed to meet performance goals and objectives.

Voting fusion is illustrated in this chapter with a three-sensor system whose detection modes involve two or more sensors. Single sensor detection modes are not implemented to avoid the shortcomings of the parallel sensor output

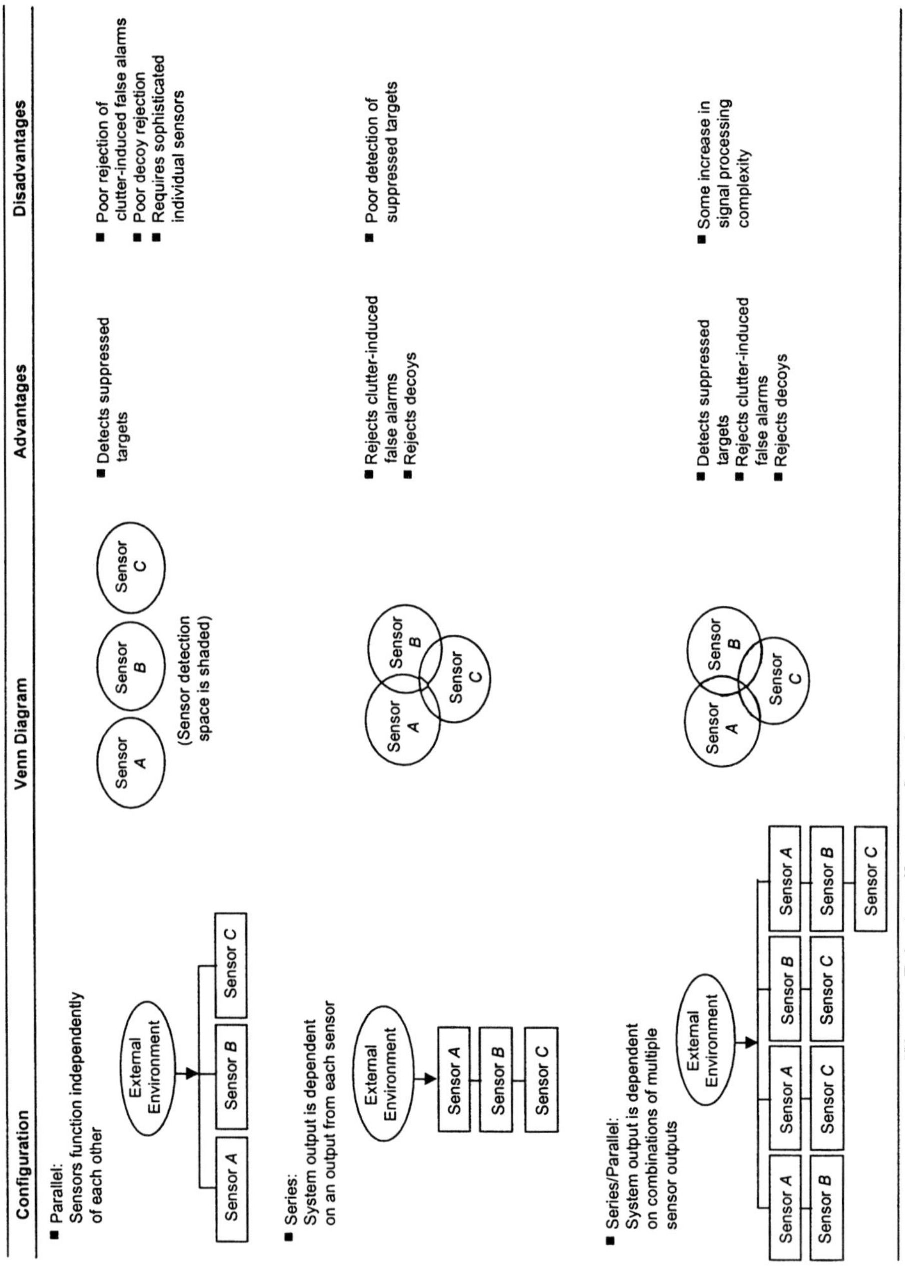

Figure 7.1: Attributes of series and parallel sensor output combinations

configuration. The sensors are assumed to operate using sensor-level fusion, where fully processed sensor data are sent to the fusion processor as target reports that contain the object detection or classification decision, associated confidence, and object location information.

In general, the fusion algorithm combines the target report data from all the sensors to assess the identity of the potential target, its track, and the immediacy of the threat. In the classification application discussed here, the voting algorithm, based on Boolean algebra, gives closed-form expressions for the multiple sensor system's estimation of the true target detection probability and false alarm probability. In order to correlate the confidence levels with the detection and false alarm probabilities, the characteristics of the sensor input signals (such as spatial frequency, bandwidth, and minimum detectable signal) and the features in the signal processing algorithms used to compare the input signals to those of known targets must be well understood. The procedures for relating confidence levels to detection and false alarm probabilities are described in this chapter through application examples.

7.1 Sensor target reports

Detection information contained in the target reports reflects the degree to which the sensor input signals conform to target features, that is, the "confidence" with which the potential target or object of interest has been recognized. The features are a function of the target size, sensor operation (active or passive), and sensor design parameters such as center frequency, resolution, bandwidth, sensitivity, and other parameters that were shown in Table 3.5, as well as the signal processing employed. Time domain processing, for example, may use features such as amplitude, pulse width, and amplitude/width ratio. Frequency domain processing may use separation between spectral peaks, widths of spectral features, and number of scattering centers producing a return signal greater than a clutter-adaptive running-average threshold.[1] Multiple pixel, infrared radiometer or FLIR sensor imagery may employ target discriminants such as image-fill criteria where the number of pixels above some threshold is compared to the total number of pixels within the image boundaries, length/width ratio of the image, central moments, and temperature gradients across object boundaries. Target reports also contain information giving target or object location. The target can, of course, be generalized to include the recognition of decoys, jammers, regions of high clutter, and anything of interest that can be ascertained within the design attribute limits of the sensor hardware and signal processing algorithms.

7.2 Sensor detection space

Sensor system detection probability is based on combinations of sensor outputs that represent the number and the degree to which the postulated target features are matched by the input signals to the individual sensors. The sensor combinations that make up the detection space are determined by the number of sensors in the sensor suite, the resolution and algorithms used by the sensors, and the manner in which the sensor outputs are combined. These considerations are discussed below.

7.2.1 Venn diagram representation of detection space

Detection space (or classification space) of a three-sensor system having Sensors *A*, *B*, and *C* is represented by a Venn diagram in Figure 7.2. Regions are labeled to show the space associated with one-sensor, two-sensor, and three-sensor combinations of outputs.

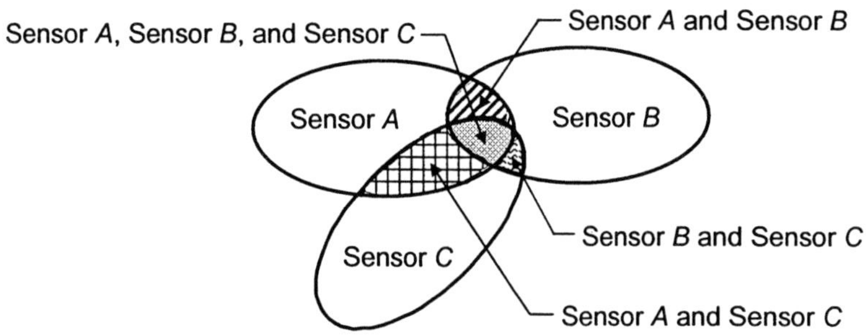

Figure 7.2: Detection modes for a three-sensor system

7.2.2 Confidence levels

Sensor detection space is not the same as confidence-level space in general, and a mapping of one into the other must be established. Nonnested or disjoint confidence levels, illustrated in the Venn diagram of Figure 7.3, are distinguished by any combination of the following that are used to define the confidence level:

- Number of signal processing algorithms or features,
- Degree of matching of the input signal to the features of an ideal target, or
- Signal-to-interference ratio.

Signal-to-interference ratio is used as a generalization of signal-to-noise ratio so that clutter can be incorporated as the limiting interference when appropriate. Nonnested confidence levels allow false alarm probabilities to be optimized for each sensor's confidence levels. This is possible because the confidence levels have a null-set intersection as described in Section 7.3.1. In the nomenclature used here, A_3 in Sensor *A* is a higher confidence level than A_2, and A_2 represents a higher confidence than A_1. Similar definitions also apply to the confidence levels of Sensors *B* and *C*.

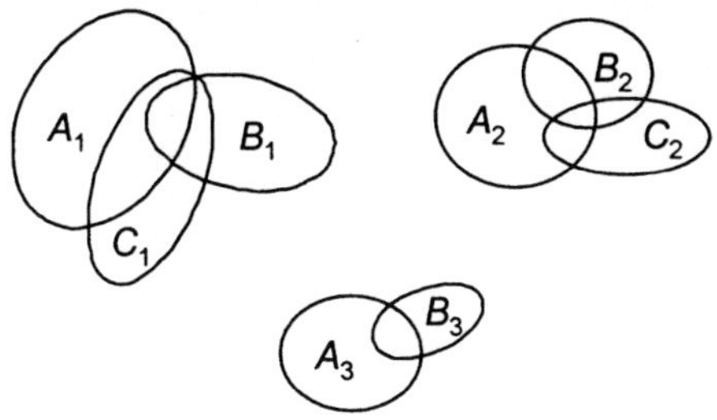

Figure 7.3: Nonnested sensor confidence levels

The number of confidence levels required of a sensor is a function of the number of sensors in the system and the ease with which it is possible to correlate target recognition features extracted from the sensor data with distinct confidence levels. The more confidence levels that are available, the easier it is to develop combinations of detection modes that meet system detection and false alarm probability requirements under wide-ranging operating conditions. Conversely, as the number of confidence levels is increased, it may become more difficult to define a set of features that unambiguously correlates a detection with a confidence level. For example, processing of radar signals in some instances contains tens of features against which the input signal is compared. Confidence levels, in this case, can reflect the number of feature matches and the degree to which the input signal conforms to the ideal target features.[2]

7.2.3 Detection modes

Combinations of sensor outputs, called detection modes, that are allowed to declare valid targets are based on the ability of the sensor hardware and signal processing discriminants to distinguish between true and false targets or countermeasure effects. Ultimately the specification of the sensor output combinations is based on the experience and knowledge of the designer and on the analysis of data gathered with the sensor system.

Allowable detection modes for the illustrative three-sensor system are given in Table 7.1. Modes that contain at least two sensors are used to avoid susceptibility to single-sensor false alarm events or countermeasures. The three-sensor mode {ABC} results from a combination of at least low confidence outputs from all sensors. The low confidence suffices because all three sensors participate in the decision. This produces a low likelihood that a false target or countermeasure-induced event will be detected as a true target, especially if the sensors respond to data that are generated from different phenomena.

Three two-sensor detection modes are also shown. The {AC} and {BC} modes use intermediate confidence levels from each of two sensors. The confidence level required has been raised, as compared to the three-sensor mode, since only two sensors are involved in making the detection decision. In mode {AB}, it is assumed that the hardware and algorithms contributing information are not as robust as they are in the other two-sensor modes. Thus, the highest third-level confidence output is required of the *A* and *B* sensors before a detection decision is made using this mode.

Table 7.1: Multiple sensor detection modes for a three-sensor system

Mode	⟵ Sensor and Confidence Level ⟶		
	A	***B***	***C***
ABC	A_1	B_1	C_1
AC	A_2	–	C_2
BC	–	B_2	C_2
AB	A_3	B_3	–

The designer may also decide that certain detection modes should be excluded altogether from the decision matrix. For example, two of the sensors may be known to false alarm on similar types of terrain. Therefore, the detection mode that results from the combination of these two sensors does not give information based on independent signature-generation phenomena and is excluded. However, these sensors, when used with a third sensor, may provide powerful target discriminants and so are retained in the sensor suite.

7.3 System detection probability

The remaining steps for calculating the system detection probability are discussed in this section. These are: derivation of the system detection probability equation based on the confidence-level structure and the selected detection mode's relation of confidence levels to detection and false alarm probabilities, computation of signal-to-noise or signal-to-clutter ratio for each sensor, and identification of the target fluctuation characteristics as observed by each sensor.

7.3.1 Derivation of system detection and false alarm probability for nonnested confidence levels

Once the detection modes are identified, Boolean algebra may be used to derive an expression for the sensor system detection probability and false alarm probability. For the above example containing one three-sensor and three two-sensor modes, the system detection probability equation takes the form

$$\text{System } P_d = P_d\{A_1\, B_1\, C_1 \text{ or } A_2\, C_2 \text{ or } B_2\, C_2 \text{ or } A_3\, B_3\}. \tag{7-1}$$

By repeated application of the Boolean algebra expansion given by

$$P\{X \text{ or } Y\} = P\{X\} + P\{Y\} - P\{XY\}, \tag{7-2}$$

Eq. (7-1) can be expanded into a total of fifteen sum and difference terms as

$$\begin{aligned} \text{System } P_d = {} & P_d\{A_1\, B_1\, C_1\} + P_d\{A_2\, C_2\} + P_d\{B_2\, C_2\} + P_d\{A_3\, B_3\} \\ & - P_d\{B_2\, C_2\, A_3\, B_3\} - P_d\{A_2\, C_2\, B_2\} - P_d\{A_2\, C_2\, A_3\, B_3\} \\ & + P_d\{A_2\, C_2\, B_2\, A_3\, B_3\} - P_d\{A_1\, B_1\, C_1\, A_2\, C_2\} \\ & - P_d\{A_1\, B_1\, C_1\, B_2\, C_2\} - P_d\{A_1\, B_1\, C_1\, A_3\, C_3\} \\ & + P_d\{A_1\, B_1\, C_1\, B_2\, C_2\, A_3\, B_3\} + P_d\{A_1\, B_1\, C_1\, A_2\, C_2\, B_2\} \\ & + P_d\{A_1\, B_1\, C_1\, A_2\, C_2\, A_3\, B_3\} \\ & - P_d\{A_1\, B_1\, C_1\, A_2\, B_2\, C_2\, A_3\, B_3\}. \end{aligned} \tag{7-3}$$

Since the confidence levels for each sensor are independent of one another (by the nonnested or disjoint assumption), the applicable union and intersection relations are

$$P_d\{A_1 \cup A_2\} = P_d\{A_1\} + P_d\{A_2\} \tag{7-4}$$

and

$$P_d\{A_1 \cap A_2\} = 0, \tag{7-5}$$

respectively. Analogous statements apply for the other sensors.

The above relations allow Eq. (7-3) to be simplified to

$$\begin{aligned} \text{System } P_d = {} & P_d\{A_1\, B_1\, C_1\} + P_d\{A_2\, C_2\} + P_d\{B_2\, C_2\} + P_d\{A_3\, B_3\} \\ & - P_d\{A_2\, B_2\, C_2\}. \end{aligned} \tag{7-6}$$

The four positive terms correspond to each of the detection modes, while the one negative term eliminates double counting of the $\{A_2\ B_2\ C_2\}$ intersection that occurs in both $\{A_2\ C_2\}$ and $\{B_2\ C_2\}$.

If the individual sensors respond to independent signature-generation phenomena (e.g., backscatter of transmitted energy and emission of energy by a warm object) such that the sensor detection probabilities are independent of one another, then the individual sensor probabilities can be multiplied together to give

$$\begin{aligned} \text{System } P_d = {} & P_d\{A_1\}\, P_d\{B_1\}\, P_d\{C_1\} + P_d\{A_2\}\, P_d\{C_2\} + P_d\{B_2\}\, P_d\{C_2\} \\ & + P_d\{A_3\}\, P_d\{B_3\} - P_d\{A_2\}\, P_d\{B_2\}\, P_d\{C_2\}. \end{aligned} \tag{7-7}$$

The interpretation of the terms in Eq. (7-7) is explained by referring to the first term $P_d\{A_1\}\ P_d\{B_1\}\ P_d\{C_1\}$. The factors in this term show the multiplication of the detection probability associated with confidence level 1 of Sensor *A* by the detection probability associated with confidence level 1 of Sensor *B* by the detection probability associated with confidence level 1 of Sensor *C*. Similar explanations may be written for the other four terms.

Equation (7-7) is also used to calculate the false alarm probability of the sensor system by replacing the detection probability by the appropriate false alarm probability for the sensor confidence level. Thus

$$\begin{aligned}\text{System } P_{fa} = {} & P_{fa}\{A_1\}\ P_{fa}\{B_1\}\ P_{fa}\{C_1\} + P_{fa}\{A_2\}\ P_{fa}\{C_2\} \\ & + P_{fa}\{B_2\}\ P_{fa}\{C_2\} + P_{fa}\{A_3\}\ P_{fa}\{B_3\} \\ & - P_{fa}\{A_2\}\ P_{fa}\{B_2\}\ P_{fa}\{C_2\}.\end{aligned} \tag{7-8}$$

7.3.2 Relation of confidence levels to detection and false alarm probabilities

Mapping of the confidence-level space into the sensor detection space is accomplished by multiplying the inherent detection probability of the sensor by the conditional probability that a particular confidence level is satisfied given a detection by the sensor. Since the signal-to-interference ratio can differ at each confidence level, the inherent detection probability of the sensor can also be different at each confidence level. Thus the probability $P_d\{A_n\}$ that Sensor *A* will detect a target with confidence level A_n is

$$P_d\{A_n\} = P_d'\{A_n\}\ P\{A_n/\text{detection}\}, \tag{7-9}$$

where

$P_d'\{A_n\}$ = inherent detection probability calculated for confidence level *n* of Sensor *A* using the applicable signal-to-interference ratio, false alarm probability, target fluctuation characteristics, and number of samples integrated

and

$P\{A_n/\text{detection}\}$ = probability that detection with confidence level A_n occurs given a detection by Sensor *A*.

Similar definitions apply to the detection probabilities at the confidence levels associated with the other sensors.

Analogous relations allow the false alarm probability to be calculated at each confidence level of the sensors. Thus the probability $P_{fa}\{A_n\}$ that a detection at confidence level A_n in Sensor *A* represents a false alarm is

$$P_{fa}\{A_n\} = P_{fa}'\{A_n\}\ P\{A_n/\text{detection}\}, \tag{7-10}$$

where

$P_{fa}'\{A_n\}$ = inherent false alarm probability selected for confidence level n of Sensor A

and

$P\{A_n/\text{detection}\}$ is the same as defined above.

The same value of the conditional probability factor is used to convert from confidence-level space into probability space when calculating both the detection and false alarm probabilities associated with a detection by a sensor at a particular confidence level. Other models (such as the nested confidence level example in Appendix B) that incorporate the conditional probability that a false alarm at confidence level A_n occurs, given a false alarm by Sensor A, may also be developed. The false alarm probabilities that characterize the sensor system and the confidence levels are dependent on the thresholds that establish the false alarm probabilities. However, the detection probability is a function of the false alarm probability increased by a factor equal to the signal processing gain. Signal processing gain is proportional to how well the signal matches target-like features designed into an algorithm and is related to the conditional probability factor in Eq. (7-9).

7.3.3 Evaluation of conditional probability

Conditional probabilities $P\{A_n/\text{detection}\}$ are evaluated using an off-line experiment to determine the performance of the signal processing algorithm. Target and nontarget data are processed by a trial set of algorithms containing confidence-level definitions based on the criteria discussed in Section 7.2. The number of detections passing each confidence level's criteria is noted and the conditional probabilities are then computed from these results. For example, if 1000 out of 1000 detections pass confidence level 1, then the probability is one that detection with confidence level 1 occurs, given a detection by the sensor. If 600 out of the 1000 detections pass confidence level 2, then the probability is 0.6 that detection with confidence level 2 occurs, given a detection by the sensor.

Once the conditional probabilities are established, the system detection and false alarm probabilities are computed using Eqs. (7-7) through (7-10). The first step in this procedure is to find the probability of a false alarm by the sensor at a particular confidence level using Eq. (7-10). Next, the false alarm probability of the mode is calculated by multiplying together the false alarm probabilities of the sensors at the confidence level at which they operate in the detection mode. Finally, the overall system false alarm probability is found by substituting the modal probabilities and the value for the negatively signed term into Eq. (7-8). If the system false alarm requirement is met, the algorithm contains the proper confidence level discrimination. If the requirement is not satisfied, then another choice of conditional probabilities is selected and the algorithm is adjusted to provide the new level of discrimination. The inherent sensor false alarm probabilities may also be adjusted to meet the system requirement, as explained in the following section.

7.3.4 False alarm probability

False alarm probabilities corresponding to each sensor's confidence levels can be different from one another because of the null set intersection described by Eq. (7-5). It is this characteristic that also allows the signal-to-interference ratio to differ at each confidence level. The inherent false alarm probabilities $P_{fa}'\{\bullet\}$ at each sensor's confidence levels are selected as large as possible consistent with satisfying the system false alarm requirement. This maximizes the detection probability for each mode. The resulting probability $P_{fa}\{\bullet\}$ that the sensor will false alarm when a detection occurs at the given confidence level is also dependent on the algorithm performance through the conditional probability factor in Eq. (7-10).

Two methods may be used to establish the inherent false alarm probability at each sensor confidence level. In the first, the inherent false alarm probability is made identical at all confidence levels by using the same detection threshold at all levels of confidence. The inherent detection probabilities are a function of this threshold. Although the threshold is the same at each confidence level, the detection probabilities can have different values at the confidence levels if the signal-to-interference ratios differ. Likewise, when the detection thresholds are the same at each confidence level, the false alarms can be reduced at the higher confidence levels through the subsequent benefits of the signal processing algorithms. This reduction in false alarms is modeled by multiplying the inherent false alarm probability by the conditional probability factor characteristic of the confidence level.

In the second method, the inherent false alarm probability at each confidence level is controlled by a different threshold. Higher confidence levels have higher thresholds. False alarms are also reduced by subsequent signal processing as above. In this case, the detection probability is a function of the different thresholds and, hence, the different false alarm probabilities that are associated with the confidence levels.

Either method may be employed to control false alarms. The off-line experiment will have to be repeated, however, to find new values for the conditional probabilities if the false alarm control methodology is changed.

In any detection mode there is a choice in how to distribute the false alarm probabilities among the different sensors. The allocations are based on the ability of the sensor's anticipated signal processing to reject false alarms, and ultimately on the conditional probabilities that relate inherent false alarm probability to the probability that the sensor will false alarm when a detection occurs at the particular confidence level. The tradeoff between conditional probability and low false alarm and detection probabilities becomes obvious from Eq. (7-10). It can be seen that as the conditional probability for any confidence level is decreased to reduce false alarms, the corresponding detection probability also decreases.

7.3.5 Detection probability

The final steps in calculating the system detection probability require the use of target, background, and sensor models to compute the signal-to-clutter or noise ratios and number of samples integrated. Upon deciding on the fluctuation characteristics that apply to the target, the inherent detection probabilities for each confidence level are calculated or found in a table or figure corresponding to the active (microwave, MMW, or laser radar) or passive (IR or MMW radiometer, FLIR, or IRST) sensor type and the direct (sensor does not contain a mixer to translate the frequency of the received signals) or heterodyne (sensor contains a mixer) detection criterion.[3,4,5,6] The probability of a detection by the sensor at a particular confidence level is found by multiplying the inherent detection probability by the conditional probability. Then the modal detection probability is obtained by multiplying together the sensor detection probabilities corresponding to the confidence levels in the detection mode. Finally, the overall system detection probability is calculated by substituting the modal detection probabilities and the value for the negatively signed term into Eq. (7-7).

7.3.6 Summary of detection probability computation model

The procedure for computing the sensor system detection probability is shown in Figure 7.4. The steps are summarized below.

1. Determine allowable sensor output combinations (detection modes).
2. Select the inherent false alarm probability for each sensor's confidence levels.
3. Through an off-line experiment, determine the number of detections corresponding to the sensor confidence levels and calculate the conditional probabilities defined in Eq. (7-9).
4. Calculate the probabilities that detections at given confidence levels represent false alarms using Eq. (7-10).
5. Calculate the sensor system false alarm probability using Eq. (7-8).

6. Note inherent false alarm probability at the confidence levels of each sensor.

7. Compute the signal-to-clutter, signal-to-noise, or signal-to-clutter plus noise ratios, as appropriate, as well as the number of samples integrated, if applicable.

8. Determine the target fluctuation characteristics that apply, e.g., steady state, slow fluctuation, and fast fluctuation.

9. Calculate the inherent sensor detection probability at each confidence level.

10. Calculate the probabilities for target detection by each sensor at the appropriate confidence levels using Eq. (7-9).

11. Calculate the sensor system detection probability using Eq. (7-7) and verify that the requirements are satisfied.

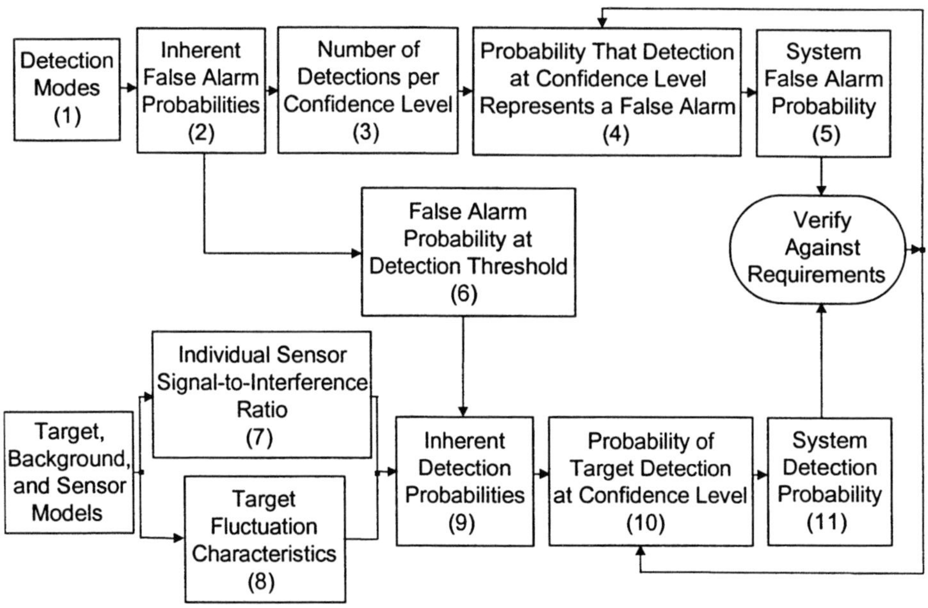

Figure 7.4: Sensor system detection probability computation model

7.4 Application example

Consider the design of a three-sensor system that must achieve a false alarm probability equal to or less than 10^{-6} with a detection probability greater than or equal to 0.8.

In this example, Sensor A is assumed to be a millimeter-wave radar to which the target has Swerling III fluctuation characteristics. Sensor B is a laser radar to which the target behaves as a Swerling II fluctuation model. Sensor C is an imaging infrared radiometer to which the target appears nonfluctuating. Different thresholds have been assumed at the sensor confidence levels.

Suppose that through an off-line experiment the number of detections corresponding to each sensor's confidence levels is determined as shown in Table 7.2. The number of detections is governed by the threshold settings, signal processing approach, and target discrimination features that are selected for each sensor. For example, based on the signal processing used in Sensor A, 600 threshold crossings out of 1000 were observed to satisfy confidence level A_2 and 400 observed to satisfy confidence level A_3. These data determine the conditional probabilities listed in Table 7.2 and subsequently used in Eqs. (7-9) and (7-10).

Table 7.2: Distribution of detections and signal-to-noise ratios among sensor confidence levels

Sensor	←	*A*	→	←	*B*	→	← *C*	→
Confidence level	A_1	A_2	A_3	B_1	B_2	B_3	C_1	C_2
Distribution of detections	1000	600	400	1000	500	300	1000	500
Conditional probability	1.0	0.6	0.4	1.0	0.5	0.3	1.0	0.5
Signal-to-noise ratio (dB)	10	13	16	14	17	20	11	15

7.4.1 Satisfying the false alarm probability requirement

False alarm probabilities are chosen as large as possible, consistent with satisfying the system false alarm requirements, in order to maximize system detection probability. With the selections shown in Table 7.3 for $P_{fa}'\{\bullet\}$ and the conditional probability data from Table 7.2, the system false alarm probability is calculated from Eqs. (7-8) and (7-10) as

$$\begin{aligned}\text{System } P_{fa} &= 2.6 \times 10^{-7} + 2.4 \times 10^{-7} + 2.5 \times 10^{-7} + 2.4 \times 10^{-7} - 2.4 \times 10^{-10} \\ &= 9.9 \times 10^{-7}, \end{aligned} \tag{7-11}$$

which satisfies the requirement of 10^{-6} or less.

Table 7.3: Inherent false alarm probabilities at the confidence levels and detection modes of the three-sensor system

Mode	Sensor *A*	Sensor *B*	Sensor *C*	Mode P_{fa}
$A_1\ B_1\ C_1$	$1.6\times10^{-2}\times1.0 = 1.6\times10^{-2}$	$1.6\times10^{-2}\times1.0 = 1.6\times10^{-2}$	$1\times10^{-3}\times1.0 = 1.0\times10^{-3}$	2.6×10^{-7}
$A_2\ C_2$	$1.6\times10^{-3}\times0.6 = 9.6\times10^{-4}$	—	$5\times10^{-4}\times0.5 = 2.5\times10^{-4}$	2.4×10^{-7}
$B_2\ C_2$	—	$2.0\times10^{-3}\times0.5 = 1.0\times10^{-3}$	$5\times10^{-4}\times0.5 = 2.5\times10^{-4}$	2.5×10^{-7}
$A_3\ B_3$	$1.2\times10^{-3}\times0.4 = 4.8\times10^{-4}$	$1.7\times10^{-3}\times0.3 = 5.1\times10^{-4}$	—	2.4×10^{-7}

7.4.2 Satisfying the detection probability requirement

Sensor detection probability at each confidence level is calculated from the inherent false alarm probability at the confidence level, signal-to-noise ratio, number of samples integrated, and appropriate target fluctuation characteristics. The selected signal-to-noise ratios and corresponding false alarm probabilities require only one sample per integration time to satisfy the system detection probability requirement in this example. Noise is used as the limiting interference to simplify the calculations. The different signal-to-noise ratios at each sensor's confidence levels, shown in Table 7.2, have been postulated to aid in defining the criteria that denote the confidence levels.

Resulting detection probabilities are shown in the matrix of Table 7.4. The first entry at each confidence level is the inherent false alarm probability (in parentheses) that establishes the threshold from which the inherent sensor detection probability is found. The second entry shows the results of the detection probability calculation for the confidence level.

The sensor system detection probability is calculated by inserting the individual sensor detection probabilities for the appropriate confidence levels into Eq. (7-7). Thus,

$$\text{System } P_d = 0.39 + 0.24 + 0.21 + 0.11 - 0.11 = 0.84, \tag{7-12}$$

assuming independence of sensor detection probabilities. The first four terms represent the detection probabilities of each of the four detection modes, while the last term represents the detection probability associated with $\{A_2\ B_2\ C_2\}$. As noted earlier, this term is incorporated twice in the sum operations and, therefore, has to be subtracted to get the correct system detection probability.

Table 7.4: Detection probabilities for the confidence levels and detection modes of the three-sensor system

Mode	Sensor *A*	Sensor *B*	Sensor *C*	Mode P_d
$A_1\ B_1\ C_1$	(1.6×10^{-2}) $0.80\times1.0 = 0.80$	(1.6×10^{-2}) $0.91\times1.0 = 0.91$	(1.0×10^{-3}) 0.53×1.0 $= 0.53$	0.39
$A_2\ C_2$	(1.6×10^{-3}) $0.85\times0.60 = 0.51$	—	(5.0×10^{-4}) $0.96\times0.50 = 0.48$	0.24
$B_2\ C_2$	—	(2.0×10^{-3}) $0.88\times0.50 = 0.44$	(5.0×10^{-4}) $0.96\times0.50 = 0.48$	0.21
$A_3\ B_3$	(1.2×10^{-3}) $0.95\times0.40 = 0.38$	(1.7×10^{-3}) $0.94\times0.30 = 0.28$	—	0.11

Table entry key: Each cell represents a confidence-level entry. Inherent false alarm probability (in parentheses) is shown on the top line of a cell. Detection probability is shown on the bottom line of a cell.

Thus, the system detection probability requirement of 0.8 or greater as well as the false alarm probability requirement of 10^{-6} or less have been satisfied. If the requirements had not been met, another choice of conditional probabilities, inherent sensor false alarm probabilities, or number of samples integrated would be selected. Once the system false alarm and detection probability requirements are satisfied, the sensor hardware or signal processing algorithms are modified to meet the new levels of discrimination.

7.4.3 Observations

The use of multiple confidence levels produces detection modes with different false alarm probabilities, as well as detection probabilities. The relatively large detection probability of the $\{A_1\ B_1\ C_1\}$ mode is achieved with relatively large false alarm probabilities, i.e., 1.6×10^{-2} at confidence level 1 of Sensors *A* and *B*, and 1.0×10^{-3} at confidence level 1 of Sensor *C*. Although the smaller false alarm probabilities of the two-sensor modes reduce their detection probabilities, they do allow these modes to function optimally in the overall fusion process and contribute to the larger system detection probability. If only one confidence level were available for each sensor, the system detection probability would not be as large or the false alarm probability would not be as small.

Another interesting observation is the correspondence of the system detection and false alarm probabilities given by Eqs. (7-7) and (7-8). The fusion process increases the detection probability over that of a single-mode multiple sensor suite (e.g., 0.84 for the fusion system as compared to 0.39 for the $\{A_1\ B_1\ C_1\}$ mode). This is exactly compensated for by an increase in system false alarm probability (9.9×10^{-7} for the fusion system as compared to 2.6×10^{-7} for the $\{A_1\ B_1\ C_1\}$ mode).

7.5 Hardware implementation of sensor fusion voting logic

Figure 7.5 illustrates how AND and OR gates connected to the confidence level output states of each sensor give the required Boolean result for the system detection probability expressed by Eq. (7-7). When each sensor's confidence level is satisfied, a binary bit is set. Then when all the bits for any AND gate are set, the output of the AND gate triggers the OR gate and a validated target command is issued.

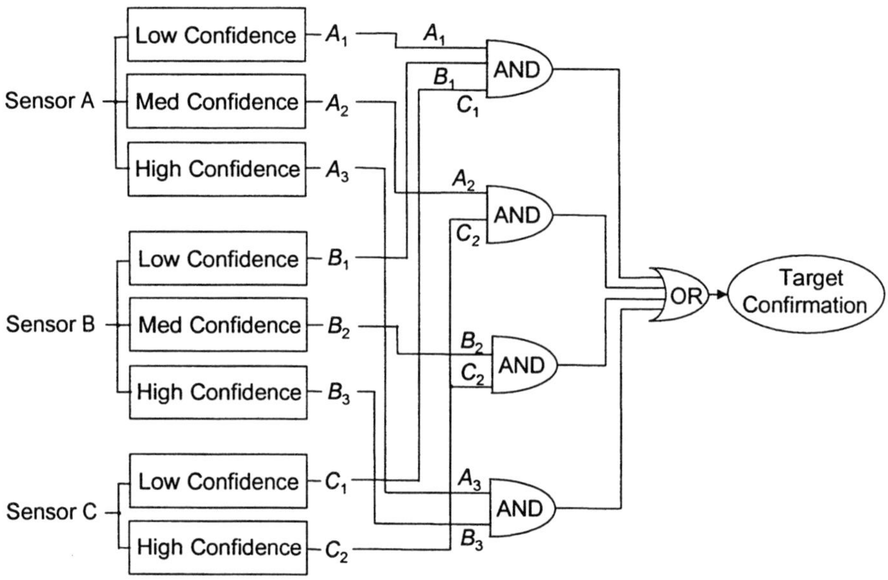

Figure 7.5: Hardware implementation for three-sensor voting fusion

For example, the $\{A_1\ B_1\ C_1\}$ mode is implemented by connecting the lowest confidence level output from the three sensors to the same AND gate. The $\{A_2\ C_2\}$ and $\{B_2\ C_2\}$ modes are implemented by connecting the intermediate confidence-level outputs from Sensors A and C and Sensors B and C, respectively, to two other AND gates. Likewise, the $\{A_3\ B_3\}$ mode is implemented by connecting the highest confidence level outputs from Sensors A and B to the last AND gate.

7.6 Comparison of voting fusion with Dempster-Shafer inference

In voting fusion, individual sensor information is used to compute detection probabilities that are combined according to a Boolean algebra expression. The principle underlying voting fusion is the combining of logical values representing sensor confidence levels, which in turn are based on predetermined detection probabilities for an object. Since weather, terrain, and countermeasures will generally not affect sensors that respond to different signature-generation phenomena in the same manner, the sensors can report different detection probabilities for the same object.

In Dempster-Shafer, the sensor information is used to compute the amount of knowledge or probability mass associated with the proposition that an object is, or is not, of a particular type. The sensors, in this case, combine compatible knowledge of the object type, using Dempster's rule to compute the probability mass associated with the intersection (or conjunction) of the propositions in the observation space.

The probability mass is analogous to the confidence level information characteristic of the voting fusion method. However, whereas voting fusion combines the sensor confidence levels through logic gates, Dempster-Shafer combines the probability masses through Dempster's rule.

7.7 Summary

Boolean algebra has been used to derive an expression for the system detection probability of a three-sensor system operating with sensors that are sensitive to independent signature-generation phenomenologies. Detection modes consisting of combinations of two and three sensors have been proposed to provide robust performance in clutter, inclement weather, and countermeasure environments. The elimination of single-sensor target-detection modes reduces sensitivity to false targets and countermeasures and increases the likelihood that targets whose signatures are suppressed in one domain will still be detected. The sensor detection modes are defined through multiple confidence levels in each sensor.

Nonnested confidence levels allow the detection probability to be independently selected and implemented at each sensor confidence level. The false alarm probabilities corresponding to the sensor confidence levels can be established in two ways. The first uses a common threshold to define the inherent false alarm probability at all the confidence levels of a particular sensor. The second allows the detection threshold, and hence inherent false alarm probability, to differ at each confidence level. The transformation of confidence-level space into detection space is accomplished by multiplying two factors. The first factor is the inherent detection probability that characterizes the sensor confidence level. The second factor is the conditional probability that detection with that confidence level occurs given a detection by the sensor. Analogous transformations permit the false alarm probability to be calculated at the confidence levels of the sensors. The simple hardware implementation for voting fusion follows from the Boolean description of the sensor-level fusion process and leads to a low-cost implementation for the fusion algorithm.

References

1. J.R. Mayersak, "An alternate view of munition sensor fusion," *Sensor Fusion*, *Proc. SPIE* 931, 64-73 (1988).
2. L.A. Klein, "A Boolean algebra approach to multiple sensor voting fusion," *IEEE Trans. Aerospace and Electron. Sys.*, AES-29, 1-11 (April 1993).
3. J.V. DiFranco and W.L. Rubin, *Radar Detection*, Prentice-Hall, New York, NY (1968).
4. D.P. Meyer and H.A. Mayer, *Radar Target Detection*, Academic Press, New York, NY (1973).
5. *Electro-Optics Handbook 11*, Second Edition, RCA, Harrison, NJ (1974).
6. R.H. Kingston, *Detection of Optical and Infrared Radiation*, Springer-Verlag, Berlin (1978).

Chapter 8

Fuzzy Logic and Fuzzy Neural Networks

Fuzzy logic is a method of representing analog processes in a digital framework. These processes are generally not easily separated into discrete segments and may be difficult to model with conventional mathematical or rule-based paradigms that require hard boundaries or decisions, i.e., binary logic where elements are a member of a given set or they are not. Consequently, fuzzy logic is valuable where the boundaries between sets of values are not sharply defined or there is partial occurrence of an event. In fuzzy set theory, an element's membership in a set is a matter of degree.

8.1 Conditions under which fuzzy logic provides an appropriate solution

Lotfi Zadeh originally developed fuzzy set theory in 1965. Zadeh reasoned that the rigidity of conventional set theory made it impossible to account for vagueness, imprecision, and shades of gray that are commonplace in real-world events.[1,2] By establishing rules and fuzzy sets, fuzzy logic creates a control surface that allows designers to build a control system even when their understanding of the mathematical behavior of the system is incomplete. Fuzzy logic permits the incorporation of the concept of vagueness into decision theory. For example, an observer may say that a person is "short" without specifying their actual height as a number. One may postulate that a reasonable specification for an adult of short stature is anyone less than 5 feet. However, other observers may declare 5 feet-2 inches or 5 feet-3 inches the cutoff between average and short height. Other examples of vagueness abound. An object may be said to be "near" or "far" from the observer, or that an automobile is traveling "faster" than the speed limit. In these examples, there is a range of values that satisfies the subjective term in quote marks.

The conditions under which fuzzy logic is an appropriate method for providing optimum control are:

- One or more of the control variables are continuous;
- Deficiencies are present in the mathematical model of the process
  - ❑ Model does not exist;
  - ❑ Model exists but is difficult to encode;

- ❑ Model is too complex to be evaluated in real time;
- ❑ Memory demands are too large;

- High ambient noise is of concern;
- Inexpensive sensors or low-precision microcontrollers must be used;
- An expert is available to specify rules that govern the system behavior and the fuzzy sets that represent the characteristics of each variable.

8.2 Application of fuzzy logic to an automobile antilock braking system

The application of fuzzy control may be illustrated by examining an automobile antilock braking system. Here control rules are established for variables such as the auto's speed, brake pressure, brake temperature, interval between brake applications, and the angle of the auto's lateral motion relative to its forward motion. These variables are all continuous. Accordingly, the descriptor that characterizes a variable within its range of values is subject to the interpretation of the designer (e.g., speed characterized as fast, slow; pressure as high, low; temperature as cold, hot; and interval as large, small).[3]

Expanded ranges of temperature states such as cold, cool, tepid, warm, and hot are needed to fully specify the temperature variable. Yet, the change from one state to another is not precisely defined. Thus, a temperature of 280°F may belong to the warm or hot state depending on the interpretation afforded by the designer. But at no point can an increase of one-tenth of a degree be said to change a "warm" condition to one that is "hot." Therefore, the concept of cold, hot, etc. is subject to different interpretations by different experts at different points in the domain of the variable.

Fuzzy logic permits control statements to be written to accommodate the imprecise states of the variable. In the case of brake temperature, a fuzzy rule could take the form: "IF brake temperature is warm AND speed is not very fast, THEN brake pressure is slightly decreased. The degree to which the temperature is considered warm and the speed not very fast controls the degree to which the brake pressure is relaxed. In this respect, one fuzzy rule can replace many conventional mathematical rules.

8.3 Basic elements of a fuzzy system

There are three primary elements in a fuzzy system, namely, fuzzy sets, membership functions, and production rules. Fuzzy sets consist of the "imprecise-labeled" groups of the input and output variables that characterize the fuzzy system. In the antilock brake system example, the temperature variable is grouped into sets of cold, cool, tepid, warm, and hot. Each set has an associated membership function that provides a graphical representation of its boundaries.

A particular value of the variable has membership in the set between the limits of 0 and 1. Zero indicates that the value is not a member of the set, while 1 indicates that the value is completely a member of the set. Intermediate membership in a set is expressed by a number between 0 and 1, which is provided by the person defining the fuzzy sets. A variable may be a member of more than one set. In the antilock brake system, a given temperature may sometimes be a member of the warm set and at other times a member of the hot set. Thus, each member of a fuzzy set is defined by an ordered pair of numbers in which the first is the value of the variable and the second is the associated membership of the variable in one or more sets.

Bell-shaped curves were originally used to define membership functions. However, the complex calculations and similarity of results led to their replacement with triangular and trapezoidal functions. The lengths of the triangle and trapezoid bases serve as design parameters that are calibrated for satisfactory system performance. Using a heuristic model, Kosko has shown that contiguous fuzzy sets should generally overlap by approximately 25 percent in area.[4] Too much overlap may blur the distinction between the fuzzy set values. Too little overlap produces a system that resembles bivalent control, causing excessive overshoot and undershoot.

Production rules represent human knowledge in the form of "IF-THEN" logical statements. In artificial intelligence applications, IF-THEN statements are an integral part of expert systems. However, expert systems rely on binary on-off logic and probability to develop the inferences used in the production rules. Fuzzy sets incorporate vagueness into the production rules since they represent less precise linguistic terms, e.g., short, not very fast, and warm. The production rules operate in parallel and influence the output of the control system to varying degrees. The logical processing using fuzzy sets is known as fuzzy logic.

8.4 Fuzzy logic processing

The fuzzy logic computation process is outlined in Figure 8.1. The sequence of fuzzy logic processing may be divided into two broad functions – inference and defuzzification. Inference processing begins with the development of the production rules in the form of IF-THEN statements, also referred to as fuzzy associative memory. The antecedent or condition block of the rule begins with IF and the consequent or conclusion block begins with THEN. The value assigned to the consequent block is equal to the *logical product* of the activation values of the antecedent membership functions (i.e., fuzzy sets). The activation value is equal to the value of the membership function at which it is intersected by the input variable at the time instant being evaluated. If the antecedent block for a particular rule is a compound statement connected by AND, the logical product is the minimum value of the corresponding activation values of the membership functions. If the antecedent block for a particular rule is a compound statement connected by OR, the logical product is the maximum value of the activation values. All the production rules that apply to a process are evaluated simultaneously (i.e., as if linked by the OR conjunction), usually hundreds of times per second. When the logical product for the antecedents is zero, the value associated with the consequent membership function is also zero.

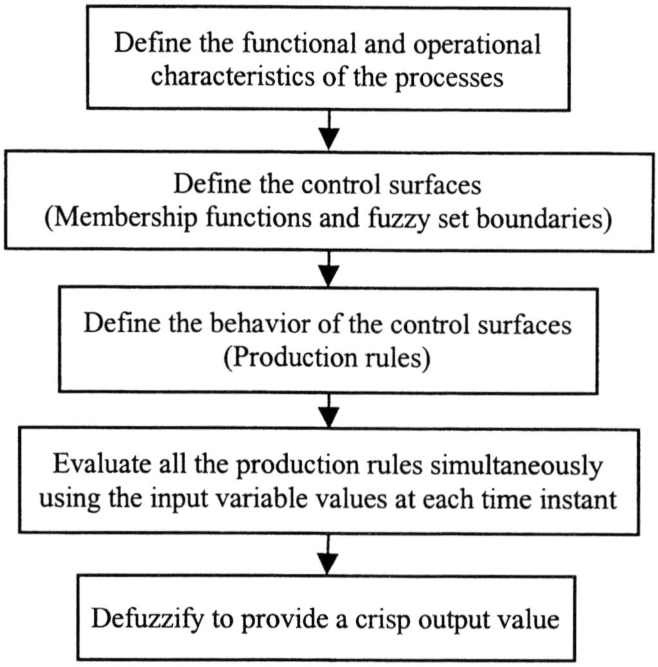

Figure 8.1: Fuzzy logic computation process

A defuzzification operation is performed to convert the fuzzy values, represented by the logical products and consequent membership functions, into a fixed and discrete output that can be utilized by the control system. Defuzzification may occur in several ways. Typically, a center of mass or fuzzy centroid computation is performed on the consequent fuzzy set. This is equivalent to finding the mode of the distribution if it is symmetric and unimodal. The fuzzy centroid incorporates all the information in the consequent fuzzy set. Two techniques are commonly used to calculate the fuzzy centroid. The first, correlation-minimum inference, clips the consequent fuzzy set at the value of the logical product as shown in Figure 8.2a. An alternative approach uses correlation-product inference, which scales the consequent fuzzy set by multiplying it by the logical product value as illustrated in Figure 8.2b. In this sense, correlation-product inferencing preserves more information than correlation-minimum inferencing.[4] Still other methods are sometimes used for defuzzification. In the simplest approach, the maximum peak is selected as the defuzzified output value. It is applicable when the distribution formed by the logical product and consequent membership functions has a unique peak.[4,5]

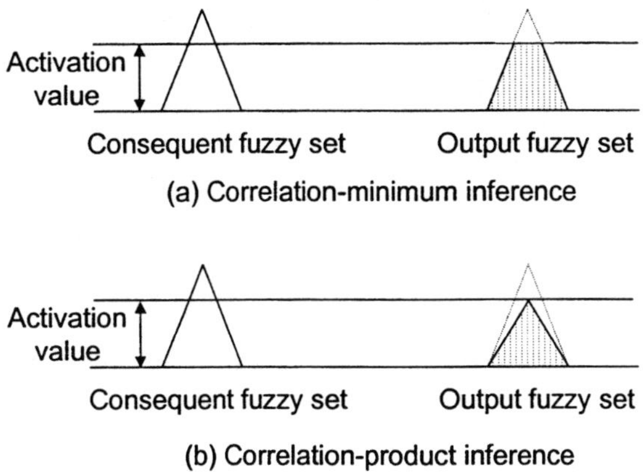

Figure 8.2: Shape of consequent membership functions for correlation-minimum and correlation-product inferencing

8.5 Fuzzy centroid calculation

Following the derivation given by Kosko, the fuzzy centroid c_k is[4]

$$c_k = \frac{\int y\, m_o(y)\, dy}{\int m_o(y)\, dy}, \tag{8-1}$$

where the limits of integration correspond to the entire universe of output parameter values,

y = output variable,

$m_o(y)$ = combined output fuzzy set formed by the simultaneous evaluation of all the production rules at time k

$$= \sum_{i=1}^{N} m_{o_i}(y), \tag{8-2}$$

N = number of production rules, and

o_i = output or consequent fuzzy set for i^{th} production rule.

When the universe of output parameter values is discretized to p values, Eq. (8-1) becomes

$$c_k = \frac{\sum_{j=1}^{p} y_j\, m_o(y_j)}{\sum_{j=1}^{p} m_o(y_j)}. \tag{8-3}$$

When the output fuzzy set is found using correlation-product inference, the global centroid c_k can be calculated from the local production rule centroids according to

$$c_k = \frac{\sum_{i=1}^{N} w_i\, c_i\, A_i}{\sum_{i=1}^{N} w_i\, A_i}, \tag{8-4}$$

where

w_i = activation value of the ith production rule's consequent set L_i,

c_i = centroid of the ith production rule's consequent set L_i

$$= \frac{\int y\, m_{L_i}(y)\, dy}{\int m_{L_i}(y)\, dy}, \tag{8-5}$$

A_i = area of the ith production rule's consequent set L_i

$$= \int m_{L_i}(y)\, dy, \tag{8-6}$$

and L is the library of consequent sets.

Furthermore, when all the output fuzzy sets are symmetric and unimodal (e.g., triangles or trapezoids) and the number of library fuzzy sets is limited to seven, then the fuzzy centroid can be computed from only seven samples of the combined output fuzzy set o. In this case,

$$c_k = \frac{\sum_{j=1}^{7} y_j\, m_o(y_j)\, A_j}{\sum_{j=1}^{7} m_o(y_j)\, A_j}, \tag{8-7}$$

where A_j is the area of the jth fuzzy-set value, equal to I_i above. Thus, Eq. (8-7) provides a simpler but equivalent form of Eq. (8-1), if all the fuzzy sets are symmetric and unimodal and if correlation-product inference is used to form the output fuzzy sets o_i.

8.6 Balancing an inverted pendulum with fuzzy logic control

A control problem often used to illustrate the application of fuzzy logic is the balance of an inverted pendulum (equivalent to the balance of a stick on the palm of a hand) as depicted[4,6] in Figure 8.3.

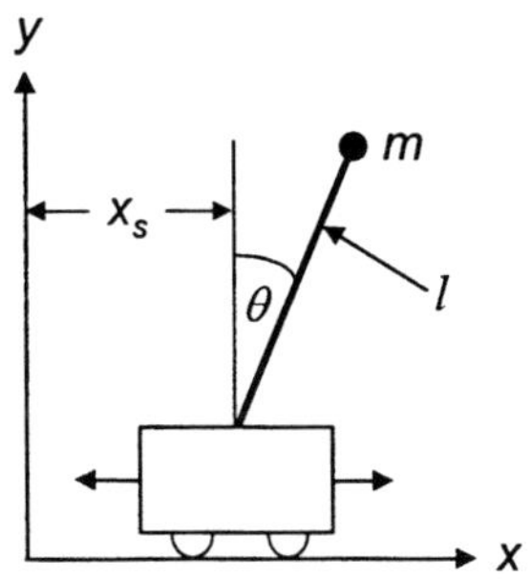

Figure 8.3: Model for balancing an inverted pendulum

8.6.1 Conventional mathematical solution

The mathematical model for a simple pendulum attached to a support driven horizontally with time is used to solve the problem with conventional control methods. The weight of the rod of length l is negligible compared to the weight of the mass m at the end of the rod in this model.

The x, y position and $\dot{x}$, $\dot{y}$ velocity coordinates of the mass m are expressed as

$$x, y = x_s + l \sin\theta, -l \cos\theta \qquad (8\text{-}8)$$

and

$$\dot{x},\ \dot{y} = \dot{x}_S + l\dot{\theta}\cos\theta,\ l\dot{\theta}\sin\theta, \qquad (8\text{-}9)$$

where θ is the angular displacement of the pendulum from equilibrium and a dot over a variable denotes differentiation with respect to time.

The equation of motion that describes the movement of the pendulum is found from the Lagrangian L of the system given by

$$L = T - V, \qquad (8\text{-}10)$$

where T is[7,8] the kinetic energy and V the potential energy of the pendulum as a function of time t. Upon substituting the expressions for the kinetic and potential energy, the Lagrangian becomes

$$L = \frac{m}{2}(\dot{x}_s^2 + l^2\dot{\theta}^2 + 2l\,\dot{x}_s\,\dot{\theta}\cos\theta) + m\,g\,l\cos\theta\,, \tag{8-11}$$

where $\dot{\theta}$ is the rate of change of angular displacement and g is the acceleration due to gravity.

The equation of motion is expressed by Lagrange's equation as[9,10]

$$\frac{d}{dt}\left(\frac{\partial L}{\partial \dot{\theta}}\right) - \frac{\partial L}{\partial \theta} = 0\,. \tag{8-12}$$

Substituting Eq. (8-11) into Eq. (8-12) gives

$$l\ddot{\theta} + \ddot{x}_s\cos\theta + g\sin\theta = 0\,. \tag{8-13}$$

The solution of Eq. (8-13) is not elementary because it involves an elliptic integral.[11] If θ is small ($|\theta| < 0.3$ rad), however, $\sin\theta$ and θ are nearly equal and Eq. (8-13) is closely approximated by the simpler equation

$$l\ddot{\theta} + \ddot{x}_s + g\theta = 0\,. \tag{8-14}$$

When $x_s = x_0\cos\omega t$, Eq. (8-14) becomes

$$\ddot{\theta} + \omega_0^2\,\theta = \frac{x_0}{l}\omega^2\cos\omega t\,, \tag{8-15}$$

where

$$\omega_0 = \sqrt{\frac{g}{l}}\,. \tag{8-16}$$

A particular solution of Eq. (8-15) obtained using the method of undetermined coefficients is[9]

$$\theta_p(t) = \frac{x_0\,\omega^2\cos\omega t}{l(\omega_0^2 - \omega^2)} \quad \text{if } \omega_0 \neq \omega. \tag{8-17}$$

The general solution of Eq. (8-15) is then

$$\theta(t) = \frac{x_0\,\omega^2 \cos\omega t}{l\,(\omega_0^2 - \omega^2)} + A\cos\omega_0 t + B\sin\omega_0 t \ \text{ for } \omega_0 \neq \omega. \tag{8-18}$$

As long as $\omega \neq \omega_0$, the motion of the pendulum is bounded. Resonance (i.e., build up of large amplitude angular displacement) occurs if $\omega_0 = \omega$. At resonance, the equation of motion becomes

$$\theta(t) = \frac{x_0\,\omega_0\, t \sin\omega_0 t}{2l} + A\cos\omega_0 t + B\sin\omega_0 t\,. \tag{8-19}$$

The constants A and B are evaluated from boundary conditions imposed on θ and $\dot{\theta}$ at $t = 0$.

8.6.2 Fuzzy logic solution

Fuzzy logic generates an approximate solution that does not require knowledge of the mathematical equations that describe the motion of the pendulum or their solution. Instead, the seven production rules listed in Table 8.1 are applied.

Table 8.1: Production rules for balancing an inverted pendulum

Rule	Antecedent Block	Consequent Block
1	IF the stick is inclined moderately to the left *and* is almost still	THEN move the hand moderately to the left quickly
2	IF the stick is inclined a little to the left *and* is falling slowly	THEN move the hand moderately to the left a little quickly
3	IF the stick is inclined a little to the left *and* is rising slowly	THEN don't move the hand much
4	IF the stick is inclined moderately to the right *and* is almost still	THEN move the hand moderately to the right quickly
5	IF the stick is inclined a little to the right *and* is falling slowly	THEN move the hand moderately to the right a little quickly
6	IF the stick is inclined a little to the right *and* is rising slowly	THEN don't move the hand much
7	IF the stick is almost not inclined *and* is almost still	THEN don't move the hand much

Production rules describe how the states of the input variables are combined. In this example, the input variables are the angle θ the pendulum makes with the vertical and the instantaneous rate of change of the angle, now denoted by $\Delta\theta$.

Both variables take on positive and negative values. The antecedent membership functions that correspond to each variable represent the ambiguous words in the antecedent block of the rules, such as "moderately," "a little," and "slowly." These words are coded into labels displayed on the membership functions shown in Figure 8.4.

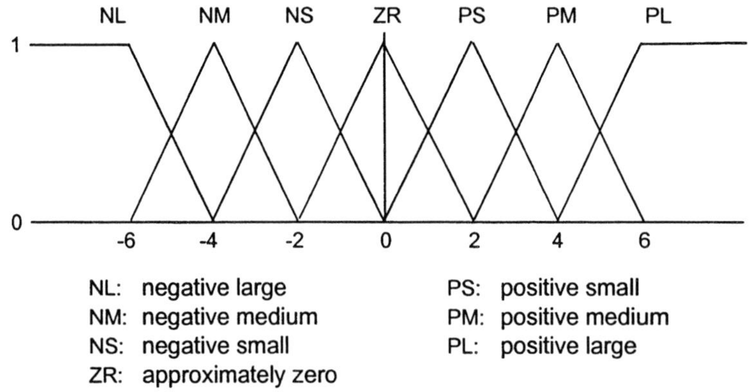

Figure 8.4: Triangle-shaped membership functions for the inverted pendulum example

The seven labels consist of three ranges in the positive direction, three in the negative direction, and a zero. The membership functions for each variable overlap by approximately 25 percent in area to ensure a smooth system response when the input level is not clear or when the level changes constantly. The membership functions describe the degree to which θ and $\Delta\theta$ belong to the fuzzy sets. The numbers at the bases of the triangular membership functions are used later to identify the centroids of each fuzzy set.

Consequent membership functions specify the motion of the pendulum base resulting from the θ and $\Delta\theta$ values input to the antecedent block. The minimum of the activation values of the antecedent membership functions is selected as the input to the consequent fuzzy sets since the antecedent conditions are linked by AND. Finally, the distribution formed by the simultaneous evaluation of all the production rules is defuzzified. In this example, a center of mass or fuzzy centroid calculation is used to compute the crisp value for the velocity of the base of the pendulum.

The fuzzy processing sequence for balancing the inverted pendulum is illustrated in Figure 8.5 for a single time instant. One input to the fuzzy controller is provided by a potentiometer that measures the angle θ. The second input represents $\Delta\theta$ as approximated by the difference between the present angle measurement and the previous angle measurement. The output of the control system is fed to a servomotor that moves the base of the pendulum at velocity Δv. If the pendulum falls to the left, its base should move to the left and vice versa.

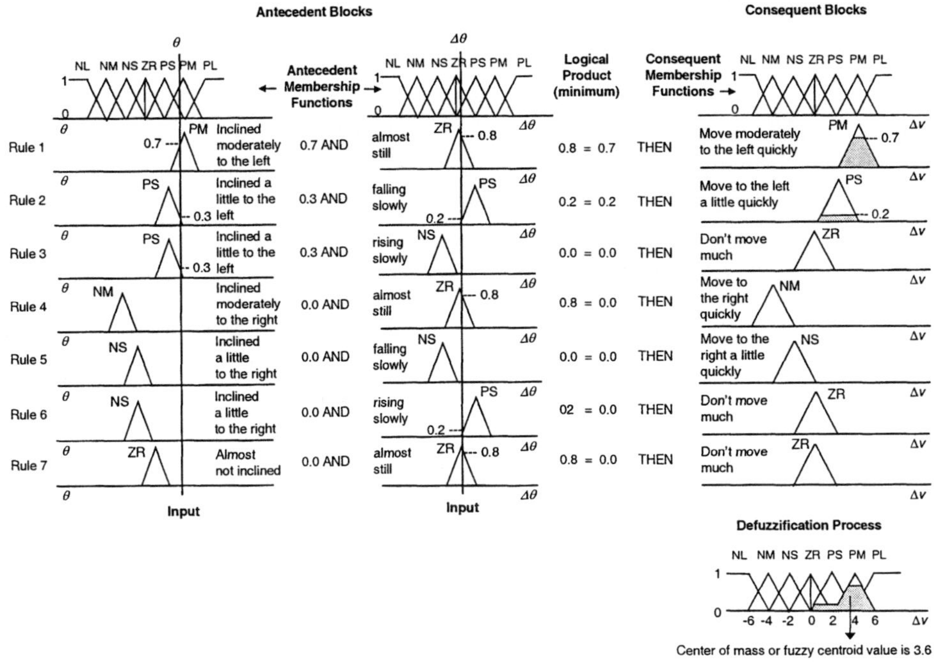

Figure 8.5: Fuzzy logic inferencing and defuzzification process for balancing an inverted pendulum

(G. Anderson, "Applying fuzzy logic in the real world," Reprinted with permission of *Sensors* magazine (www.sensorsmag.com), Sept. 1992. Helmers Publishing, Inc. Copyright 1992.

Examining the antecedent block for Production Rule 1 in Figure 8.5 shows that θ intercepts the membership function for "inclined moderately to the left" at 0.7 and $\Delta\theta$ crosses the membership function for "almost still" at 0.8. The logical product of these two values is 0.7, the minimum value of the two inputs. The value of 0.7 is next associated with the consequent block of Production Rule 1, "move moderately to the left quickly." Proceeding to Production Rule 2, we find that θ intercepts the membership function for "inclined a little to the left" at 0.3 and $\Delta\theta$ crosses the membership function for "falling slowly" at 0.2. The logical product value of 0.2 is then associated with the consequent block of Production Rule 2, "move to the left a little quickly." The logical products for the remaining production rules are zero since at least one of the antecedent membership functions is zero.

Defuzzification occurs once the simultaneous processing of all the rules is complete for the time sample. Defuzzification is performed by the center of mass calculation illustrated in the lower right corner of the figure for correlation-minimum inference. The defuzzified output controls the direction and speed of the movement required to balance the pendulum. In this case, the command instructs the servomotor to move the base of the pendulum to the left at a velocity equal to the center of mass value of 3.6.

The value of 3.6 was calculated using Eq. (8-3) and the entries in Table 8.2. The numerator in Eq. (8-3) is equal to the sum of the products of $y_j\, m_o(y_j)$, while the denominator is equal to the sum of $m_o(y_j)$ for j =1 to 7. Since the areas I_j of the consequent sets are equal, the sum of the products of w_j and c_j may be substituted for the numerator and the sum w_j for the denominator, where w_j is the activation value and c_j the centroid of the consequent of production rule j.

Table 8.2: Outputs for the inverted pendulum example

j	Consequent	w_j	c_j	w_jc_j
1	PM	0.7	4	2.8
2	PS	0.2	2	0.4
3	ZR	0	0	0
4	NM	0	-4	0
5	NS	0	-2	0
6	ZR	0	0	0
7	ZR	0	0	0
Sum		0.9		3.2

Although the output from a fuzzy system is crisp, the solution is still approximate as it is subject to the vagaries of the rule set and the membership functions. Fuzzy logic control is considered robust because of its tolerance for imprecision. Fuzzy systems can operate with reasonable performance even when data are missing or membership functions are loosely defined.

8.7 Fuzzy logic applied to multitarget tracking

This example utilizes a fuzzy Kalman filter to correct the estimate of the target position and velocity state vector at time k+1 using updated parameters at time k. The Kalman filter provides a state estimate that minimizes the mean squared error between the estimated and observed position and velocity states over the entire history of the measurements.[12,13] The discrete time implementation of the fuzzy Kalman filter reduces computation time as compared with the conventional Kalman filter implementation, especially for multidimensional, multitarget scenarios.

8.7.1 Conventional Kalman filter approach

A Kalman filter contains a state transition model and a measurement model. The *state transition model* predicts the target position and velocity coordinates of the state vector $\boldsymbol{X}$ at time k+1 based on information available at time k according to

$$\boldsymbol{X}_{k+1} = \boldsymbol{F}\,\boldsymbol{X}_k + \boldsymbol{J}\,\boldsymbol{w}_k\,, \tag{8-20}$$

where

$\boldsymbol{X}_k^{\mathrm{T}} = [x_k \;\; \dot{x}_k \;\; y_k \;\; \dot{y}_k\,]^{\mathrm{T}}$ (in two dimensions),

T = transpose operation,

$\boldsymbol{F}$ = system or transition matrix,

$\boldsymbol{J}$ = input matrix, and

$\boldsymbol{w}_k$ = disturbance input vector or process noise that exists in the target motion model.

The Kalman filter extrapolates the state vector $\boldsymbol{X}$ and measurement error covariance matrix $\boldsymbol{P}$ to time k using data collected at time k-1. Accordingly, the state vector estimate $\hat{\boldsymbol{X}}_{k|k-1}$ provided by the Kalman filter is given by

$$\hat{\boldsymbol{X}}_{k|k-1} = \boldsymbol{F}\,\hat{\boldsymbol{X}}_{k-1|k-1} \tag{8-21}$$

and the covariance matrix $\boldsymbol{P}_{k|k-1}$ by

$$\boldsymbol{P}_{k|k-1} = \boldsymbol{F}\,\boldsymbol{P}_{k-1|k-1}\,\boldsymbol{F}^{\mathrm{T}} + \boldsymbol{J}\Sigma_k\,\boldsymbol{J}^{\mathrm{T}}\,, \tag{8-22}$$

where the notation $k|k$-1 indicates the estimated or extrapolated value at time k calculated with data gathered at time k-1, and

Σ_k = covariance matrix for the disturbance input vector $\boldsymbol{w}_k$.

The *measurement model* uses new information contained in the innovation vector to correct the extrapolated state estimate. The innovation vector $\tilde{\boldsymbol{Z}}_k$ is defined as the difference between the observed and extrapolated measurement vectors such that

$$\tilde{\boldsymbol{Z}}_k = \boldsymbol{Z}_k - \hat{\boldsymbol{Z}}_{k|k-1} = \boldsymbol{Z}_k - \boldsymbol{H}\,\hat{\boldsymbol{X}}_{k|k-1}\,, \tag{8-23}$$

where

$$Z_k = H X_k + n_k \tag{8-24}$$

H = output or observation matrix, and

n_k = measurement noise vector that generally contains a fixed, but unknown bias and a random component.

Finally, the extrapolated state vector and covariance matrix in Eqs. (8-21) and (8-22) are corrected as

$$\hat{X}_{k|k} = \hat{X}_{k|k-1} + G_k \tilde{Z}_k \tag{8-25}$$

and

$$P_{k|k} = [(P_{k|k-1})^{-1} + H^{\mathrm{T}} R_k^{-1} H]^{-1}, \tag{8-26}$$

where

G_k = filter gain, a function of the measurement error covariance matrix,

$$= P_{k|k-1} H^{\mathrm{T}} (H P_{k|k-1} H^{\mathrm{T}} + R_k)^{-1} \tag{8-27}$$

and

R_k = covariance matrix of the noise vector n_k.

The corrected covariance matrix may also be written in terms of the gain as[14]

$$P_{k|k} = [I - G_k H] P_{k|k-1} \tag{8-28}$$

where I is the identity matrix. The matrix inversion lemma may be used to convert the corrected covariance matrix into the form

$$P_{k|k} = P_{k|k-1} - P_{k|k-1} H^{\mathrm{T}} [H^{\mathrm{T}} P_{k|k-1} H + R_k]^{-1} H P_{k|k-1}. \tag{8-29}$$

8.7.2 Fuzzy Kalman filter approach

Fuzzy logic reduces the time to perform complex matrix multiplications that are characteristic of higher order systems. This example treats the incomplete information case in which only the position variables are available for measurement. Fuzzy logic is used for data association and for updating the extrapolated state vector. Data are associated with a specific target by defining (1) a validation gate based on Euclidean distance and (2) a similarity measure based on object size and intensity. A fuzzy return processor is created to execute these functions. The output of this process is the average innovation vector used as the input to a fuzzy state correlator. The fuzzy state correlator updates the extrapolated state estimate of the position and velocity of the target at time k given information at time k-1.

The extrapolation equation for the state vector $\boldsymbol{X}$ that appears in the fuzzy Kalman filter is identical to Eq. (8-21). The approaches depart by using fuzzy logic to generate the correction vector $\boldsymbol{C}_k$ needed to update the state estimate according to

$$\hat{\boldsymbol{X}}_{k|k} = \hat{\boldsymbol{X}}_{k|k-1} + \boldsymbol{G}_k \, \boldsymbol{C}_k , \tag{8-30}$$

where $\boldsymbol{C}_k$ is the fuzzy equivalent of the innovation vector $\tilde{\boldsymbol{Z}}_k$.

Step 1: Fuzzy return processor — The function of the fuzzy return processor is to reduce the uncertainty in target identification caused by clutter, background noise, and image processing. In this example, the data used to identify and track the targets are produced by a sequence of forward-looking infrared (FLIR) images.[12] The passive FLIR sensor allows the position, but not the velocity, of the target to be measured. The fuzzy return processor produces two parameters that are used to associate the FLIR sensor data with a specific target. The first is based on a validation gate. The second is a similarity measure related to the rectangular size of the image and intensity of the pixels in the image.

Data validation is needed when multiple returns are received from the vicinity of the target at time k. Fuzzy validation imparts a degree of validity between 0 and 1 to each return. The validity $\beta_{valid,i}$ for the ith return is inversely related to the Euclidean norm of the innovation vector defined as

$$\left\| \tilde{\boldsymbol{Z}}_{k,i} \right\| = [(x_{k,i} - \hat{x}_k)^2 + (y_{k,i} - \hat{y}_k)^2]^{\frac{1}{2}} , \tag{8-31}$$

where

$\tilde{\boldsymbol{Z}}_{k,i}$ = innovation vector at time k for the ith return

$$= \boldsymbol{Z}_{k,i} - \hat{\boldsymbol{Z}}_{k|k-1} \text{ [analogous to Eq. (8-23)]} \tag{8-32}$$

and the parameters in parentheses represent the observed and extrapolated values of x and y, respectively.

The fuzzy membership function for the validity is illustrated in Figure 8.6. The constants d_1 and d_2 are varied to optimize the performance of the filter as the number of clutter returns changes. The degree of validity is combined with the similarity measure to calculate an average innovation vector Z_k' for use in the fuzzy state correlator.

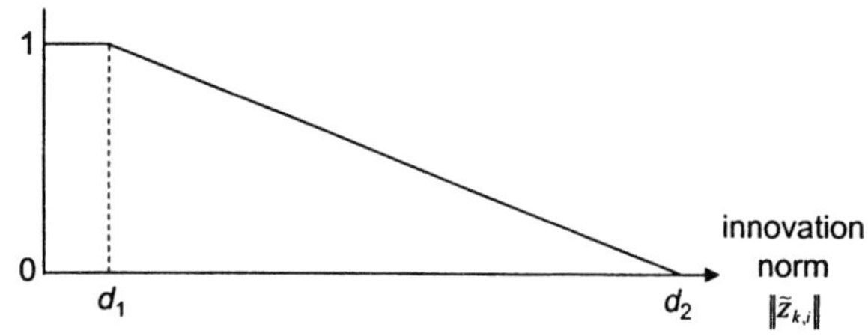

Figure 8.6: Validity membership function

The similarity measure is used to correlate new data with previously identified targets. The correlation is performed using size-difference and intensity-difference antecedent membership functions as shown in Figure 8.7.

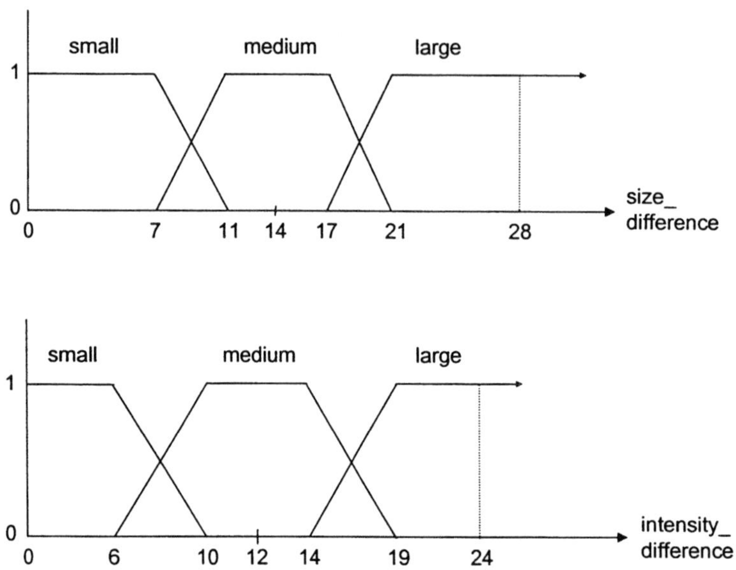

Figure 8.7: Size-difference and intensity-difference membership functions

An example of the production rules that determine if a return i falls within the size and intensity validation gate is

IF (size_diff$_i$_is_small) AND (intensity_diff$_i$_is small), THEN (degree_of_ similarity$_i$_is_high).

The complete set of production rules needed to associate new data with targets is illustrated in Table 8.3.

Table 8.3: Fuzzy associative memory rules for degree_of_similarity

Intensity_diff	Size_diff		
	Small	**Medium**	**Large**
Small	High	High	Medium
Medium	High	Medium	Low
Large	Medium	Low	Low

Once the data have been associated with previously identified targets, a similarity membership function, such as that shown in Figure 8.8, is used to find the consequent values. The result is defuzzified to find the weight $\beta_{similar,i}$ through a center of mass calculation based on the activation value of the degree_of_similarity and the inferencing method applied to the consequent fuzzy sets.

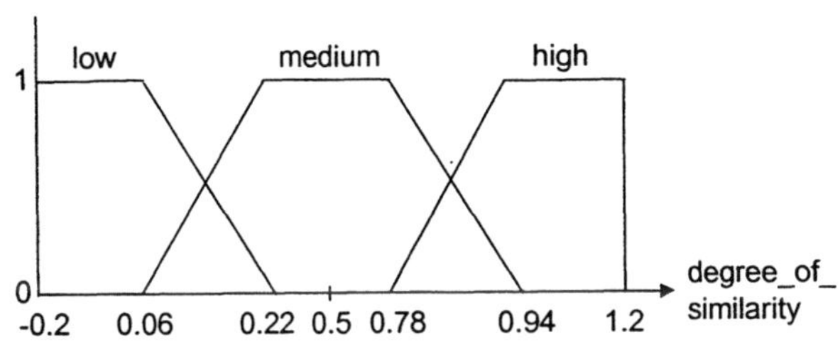

Figure 8.8: Similarity membership functions

The weights $\beta_{valid,i}$ and $\beta_{similar,i}$ found for all i = 1, ... , n returns are used to calculate a weighted average innovation vector as

$$\tilde{Z}_k' = \begin{bmatrix} \tilde{x}_k' \\ \tilde{y}_k' \end{bmatrix} = \sum_{i=1}^{n} \beta_i \tilde{Z}_{k,i} \tag{8-33}$$

where β_i, with values between 0 and 1, is the weight assigned to the ith innovation vector. It represents the belief or confidence that the identified return is the target. The value β_i is calculated as a linear combination of $\beta_{valid,i}$ and $\beta_{similar,i}$ as

$$\beta_i = b_1\, \beta_{valid,i} + b_2\, \beta_{similar,i}, \tag{8-34}$$

where the constants b_1 and b_2 sum to unity. These constants are used to alter the return processor's performance by trading off the relative importance of validity and similarity. The weighted average innovation vector as found from Eq. (8-33) is used as the input to the fuzzy state correlator for the particular target of interest.

Step 2: Fuzzy state correlator — The fuzzy state correlator calculates the correction vector $\boldsymbol{C}_k$ that updates the state estimate for the position and velocity of the target at time k according to Eq. (8-30).

To find $\boldsymbol{C}_k$, the weighted average innovation vector is first separated into its x and y components, e_x and e_y. An error vector $\boldsymbol{e}_k$ is then defined as

$$\boldsymbol{e}_k = \begin{bmatrix} e_x \\ e_y \end{bmatrix} = \tilde{\boldsymbol{Z}}_k{}'. \tag{8-35}$$

Since the x and y directions are independent, Horton and Jones[12] develop the fuzzy state correlator for the x direction and then generalize the result to include the y direction. The production rules that determine the fuzzy output of the correlator have two antecedents, the average x component of the innovation vector e_x and the differential error d_e_x. Assuming the current and previous values of the error vector, e_x and past_e_x, are available, allows d_e_x to be computed as

$$d\_e_x = (e_x - \text{past\_}e_x)/\text{timestep}. \tag{8-36}$$

The antecedent membership functions that define the fuzzy values for e_x and d_e_x are shown in Figure 8.9.

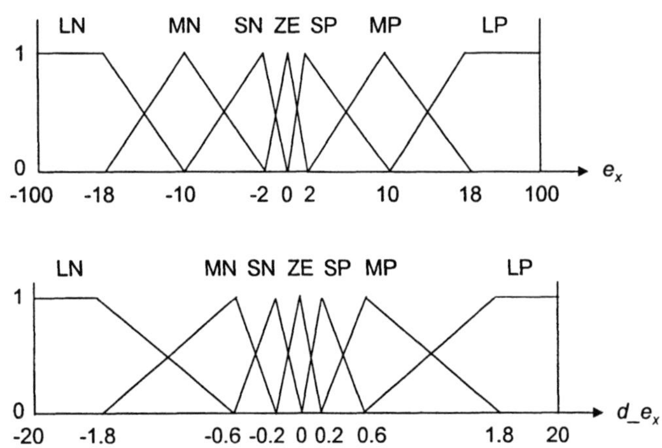

Figure 8.9: Innovation vector and the differential error antecedent membership functions

Using the values of e_x and d_e_x, the production rules for the fuzzy state correlator take the form

> IF (e_x is large negative [LN]) AND (d_e_x is large positive [LP]), THEN ($C_{k,x}$ is zero [ZE]).

Table 8.4 summarizes the 49 rules needed to implement the fuzzy state correlator.

Table 8.4: Fuzzy associative memory rules for the fuzzy state correlator

d_e_x	e_x						
	LN	**MN**	**SN**	**ZE**	**SP**	**MP**	**LP**
Large negative (LN)	LN	LN	MN	MN	MN	SN	ZE
Medium negative (MN)	LN	MN	MN	MN	SN	ZE	SP
Small negative (SN)	MN	MN	MN	SN	ZE	SP	MP
Zero (ZE)	MN	MN	SN	ZE	SP	MP	MP
Small positive (SP)	MN	SN	ZE	SP	MP	MP	MP
Medium positive (MP)	SN	ZE	SP	MP	MP	MP	LP
Large positive (LP)	ZE	SP	MP	MP	MP	LP	LP

After tuning the output to reduce the average root least-square error (RLSE), Horton and Jones find the consequent membership functions to be those shown in Figure 8.10. In this example, the bases of the trapezoidal and triangular membership functions were scaled to provide the desired system response.

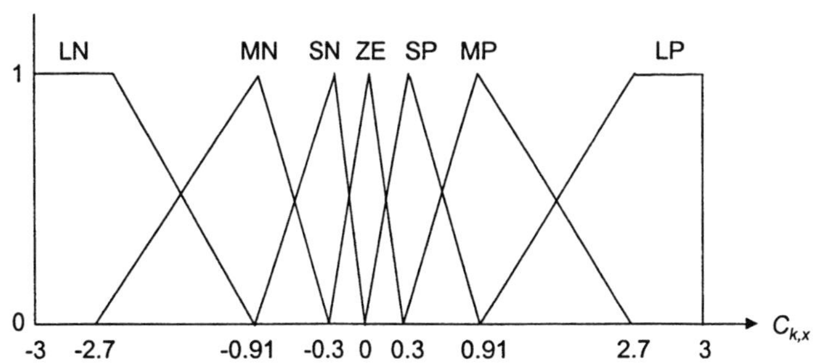

Figure 8.10: Correction vector consequent membership functions

The defuzzified output is calculated from the center of mass or fuzzy centroid corresponding to the activation value of the correction vector C_k and the inferencing method applied to the consequent fuzzy sets. The performance of the fuzzy tracker was improved by adding a variable gain Γ to the defuzzified inputs and outputs of the system as shown in Figure 8.11 for the x direction. By proper choice of gains ($\Gamma_1 = 1$, $\Gamma_2 = 1$, $\Gamma_3 = 7$), the average RLSE error was reduced to approximately 1 from its value of 5 obtained when the gains were not optimized.

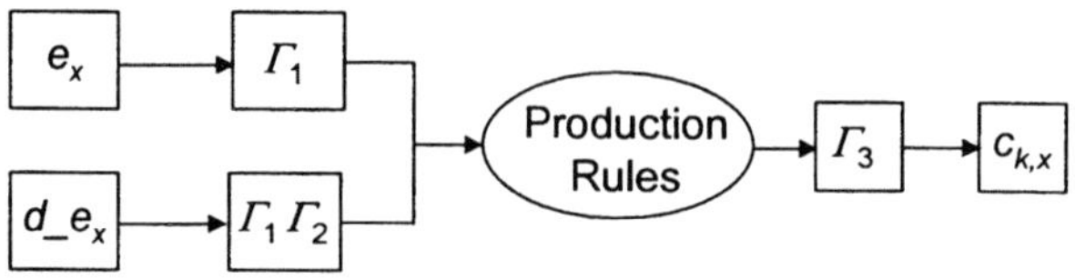

Figure 8.11: Improving performance of the fuzzy tracker by applying gains to the crisp inputs and outputs

8.8 Fuzzy neural networks

Adaptive fuzzy neural systems use sample data and neural algorithms to define the fuzzy system at each time instant. Either the weights and/or the input signals are fuzzy sets. Thus, fuzzy neural networks may be characterized by

- Real number signals with fuzzy set weights;
- Fuzzy signals with real number weights;
- Both fuzzy signals and fuzzy weights.

An example of the first class of fuzzy neural network is the fuzzy neuron developed by Yamakawa et al.[15,16] As illustrated in Figure 8.12, the neuron contains real number inputs x_i ($i = 1, \ldots, m$) and fixed fuzzy sets μ_{ik} ($k = 1, \ldots, n$) that modify the real number weights w_{ik}. The network is trained with a heuristic learning algorithm that updates the weights with a formula similar to the backpropagation algorithm. A restriction is placed on the fuzzy sets μ_{ik} such that only two neighboring μ_{ik} can be nonzero for a given x_i.

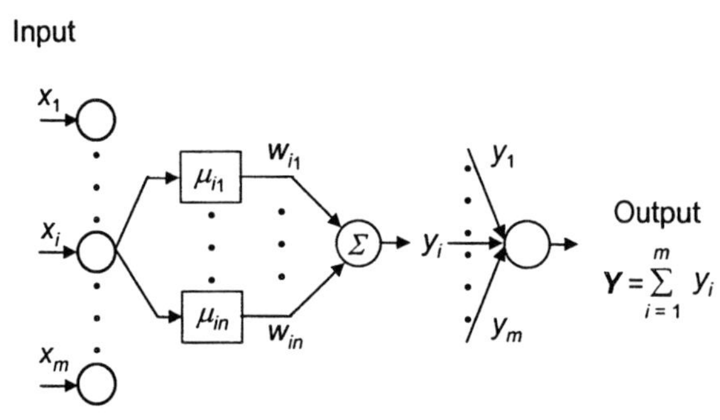

Figure 8.12: Yamakawa's fuzzy neuron

Thus, in Figure 8.12 if $\mu_{ik}(x_i)$ and $\mu_{i,k+1}(x_i)$ are nonzero, then

$$y_i = \mu_{ik}(x_i)w_{ik} + \mu_{i,k+1}(x_i)w_{i,k+1}. \quad (8\text{-}37)$$

The output Y of the neuron is equal to the sum of the y_i such that

$$Y = y_1 + y_2 + \ldots + y_m. \quad (8\text{-}38)$$

Nakamura et al.[17] and Tokunaga et al.[18] developed another type of fuzzy neuron having the topology shown in Figure 8.13.

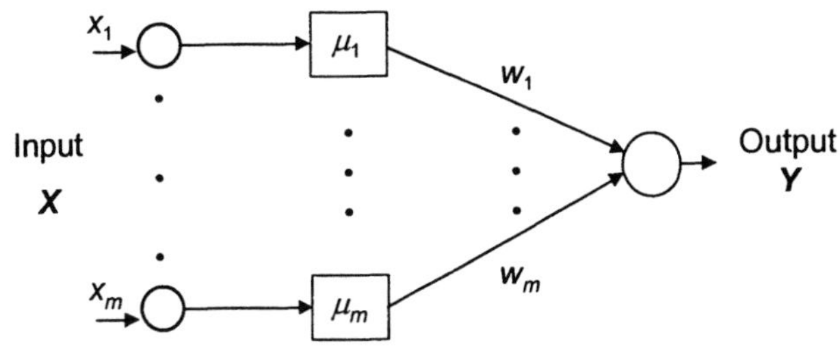

Figure 8.13: Nakamura's and Tokunaga's fuzzy neuron

The learning algorithm in this case optimizes both the trapezoidal membership functions for fuzzy sets μ_i ($i = 1, \ldots, m$) and the real number weights w_i. The output Y is equal to

$$Y = \sum_{i=1}^{m} w_i \, \mu_i(x_i) / \sum_{i=1}^{m} w_i \, . \quad (8\text{-}39)$$

The second and third classes of fuzzy neural networks are similar in topology to multilayer feedforward networks. The second class of fuzzy neural networks contains a fuzzy input signal vector and a fuzzy output signal vector. Backpropagation and other training algorithms have been proposed for this class of network.[19,20,21] The third class of fuzzy neural networks contains fuzzy input and output signal vectors and fuzzy weights that act on the signals entering each layer. Learning algorithms for the third class of fuzzy neural networks are discussed by Buckley and Hayashi.[15] They surmise that learning algorithms will probably be specialized procedures when operations other than multiplication and addition act on signals in this class of fuzzy neural networks.

8.9 Summary

Fuzzy logic, somewhat contrary to its name, is a well-defined discipline that finds application where the boundaries between sets of values are not sharply defined, where there is partial occurrence of an event, or where the specific mathematical equations that govern a process are not known. Fuzzy logic also is used to reduce the computation time that would otherwise be needed to control complex or multidimensional processes.

The value of fuzzy logic to data fusion has appeal in identifying battlefield objects, describing the composition of enemy units, and interpreting enemy intent and operational objectives. Perhaps the most difficult aspect of its application is the definition of the membership functions that specify the influence of the input variables on the fuzzy system output.

A fuzzy control system nonlinearly transforms exact or fuzzy state inputs to a fuzzy set output. Fuzzy systems contain membership functions and production rules or fuzzy associative memory. Membership functions define the boundaries of the fuzzy sets consisting of the input and output variables. The production rules operate in parallel and are activated to different degrees through the membership functions. Each rule represents ambiguous expert knowledge or learned input-output transformations. A rule can also summarize the behavior of a specific mathematical model. The output fuzzy set is usually defuzzified using a centroid calculation to generate an exact numerical output for the control system.

The balance of an inverted pendulum and track estimation with a Kalman filter were described to illustrate the wide applicability of fuzzy logic and contrast the fuzzy solution with the conventional mathematical solution. Adaptive fuzzy neural systems can also be constructed. These rely on sample data and neural algorithms to define the fuzzy system at each time instant.

References

1. L.A. Zadeh, *Fuzzy Sets and Systems*, North-Holland Press, Amsterdam (1978).
2. L.A. Zadeh, "Fuzzy logic," *IEEE Computer*, 83-93 (Apr. 1988).
3. E. Cox, "Fuzzy fundamentals," *IEEE Spectrum*, 58-61 (Oct. 1992).
4. B. Kosko, *Neural Networks and Fuzzy Systems: A Dynamical Systems Approach to Machine Intelligence*, Prentice-Hall, Englewood Cliffs, NJ (1992).
5. N. Vadiee, "Fuzzy rule-based expert systems II," Chapter 5 in *Fuzzy Logic and Control: Software and Hardware Applications*, Editors: M. Jamshidi, N. Vadiee, and T. Ross, PTR Prentice Hall, Englewood Cliffs, NJ (1993).
6. G. Anderson, "Applying fuzzy logic in the real world," *Sensors*, 15-25 (Sept. 1992).
7. R.B. Lindsay, *Physical Mechanics,* 3rd Ed., D. Van Nostrand Company, Princeton, NJ (1961).
8. L. Yung-kuo, Ed., *Problems and Solutions on Mechanics*, #2029, World Scientific, River Edge, NJ (1994).
9. H. Margenau and G.M. Murphy, *The Mathematics of Physics and Chemistry*, D. Van Nostrand Comp., Princeton, NJ (1961).
10. J. Irving and N. Mullineux, *Mathematics in Physics and Engineering*, Academic Press, New York, NY (1959).
11. E.D. Rainville, *Elementary Differential Equations*, 2nd Ed., Macmillan, New York, NY (1959).
12. M.J. Horton and R.A. Jones, "Fuzzy logic extended rule set for multitarget tracking," *Acquisition, Tracking, and Pointing IX, Proc. SPIE* 2468, 106-117 (1995).
13. Y. Bar-Shalom, K.C. Chang, and H.A.P. Blom, "Automatic track formation in clutter with a recursive algorithm," Chapter 2 in *Multitarget-Multisensor Tracking: Advanced Applications*, Editor: Y. Bar-Shalom, Artech House, Norwood, MA (1990).
14. M.P. Dana, *Introduction to Sensors and Multi-Target, Multi-Sensor Data Fusion Techniques for Detection, Identification, and Tracking, Part II*, Johns Hopkins University, Organizational Effectiveness Institute Short Course, Rockville, MD (1998).
15. J.J. Buckley and Y. Hayashi, "Fuzzy neural networks," Chapter 11 in *Fuzzy Sets, Neural Networks, and Soft Computing*, Editors: R.R. Yager and L.A. Zadeh, Van Nostrand Reinhold, New York, NY (1994).
16. T. Yamakawa, E. Uchino, T. Miki, and H. Kusanagi, "A neo fuzzy neuron and its application to system identification and prediction of the system behavior," *Proc. 2nd Int. Conf. Fuzzy Logic Neural Networks* (*IIZUKA '92*), Iizuka, Japan, July 17-22, 477-483 (1992).
17. K. Nakamura, T. Fujimaki, R. Horikawa, and Y. Ageishi, "Fuzzy network production system," *Proc. 2nd Int. Conf. Fuzzy Logic Neural Networks* (*IIZUKA '92*), Iizuka, Japan, July 17-22, 127-130 (1992).

18. M. Tokunaga, K. Kohno, Y. Hashizume, K. Hamatani, M. Watanabe, K. Nakamura, and Y. Ageishi, "Learning mechanism and an application of FFS-network reasoning system," *Proc. 2nd Int. Conf. Fuzzy Logic Neural Networks* (*IIZUKA '92*), Iizuka, Japan, 123-126 (July 17-22, 1992).
19. H. Ishibuchi, R. Fujioka, and H. Tanaka, "An architecture of neural networks for input vectors of fuzzy numbers," *Proc. IEEE Int. Conf. Fuzzy Syst.* (*FUZZ-IEEE '92*), San Diego, CA, 1293-1300 (Mar. 8-12, 1992).
20. H. Ishibuchi, H. Okada, and H. Tanaka, "Interpolation of fuzzy if-then rules by neural networks," *Proc. 2nd Int. Conf. Fuzzy Logic Neural Networks* (*IIZUKA '92*), Iizuka, Japan, 337-340 (July 17-22, 1992).
21. I. Requena and M. Delgado, "R-FN: A model of fuzzy neuron," *Proc. 2nd Int. Conf. Fuzzy Logic Neural Networks* (*IIZUKA '92*), Iizuka, Japan, 793-796 (July 17-22, 1992).

Chapter 9

Passive Data Association Techniques for Unambiguous Location of Targets

This chapter was written, in part, by Henry Heidary of Hughes Aircraft Corporation (now Raytheon Systems), Fullerton, CA.

Sensor and data fusion can occur using several types of passively acquired information. For example, the raw signals themselves or various types of information extracted from the signals may be selected as inputs to data fusion processors. The signals, sensor data, and communications media available in a particular command-and-control system often dictate the optimum data fusion technique. This chapter addresses data fusion architectures applicable to multiple sensor and multiple target scenarios in which range information to the target is missing, but where the target location is required.

Unambiguous target track files may be generated by using data association techniques to combine various types of passively acquired data from multiple sensors. In the examples described in this chapter, multiple ground-based radars are used to locate energy emitters, i.e., targets, by fusing either of three different types of received data: (1) received signal waveforms, (2) angle information expressing the direction to the emitters, or (3) emitter angle track data that are output from the sensors. The alternate fusion methods illustrate the difficulties and system design issues that arise in selecting the data fusion process and the type of passively acquired data to be fused.

These fusion techniques allow range information to be obtained from arrays of passive sensors that measure direction angles, or from active sensors where range information is denied (as for example when the sensor is jammed), or from combinations of passive and active sensors. For example, electronic support measure (ESM) radars can use the fused data to find the range to the emitters of interest. These fusion methods can also be used with surveillance radars that are jammed to locate the jammer positions. In a third application, angle data from a netted array of IRST sensors, or for that matter from acoustic or any passively operated sensors, can be fused to find the range to the emitters.

Fusion of the signal waveforms received from the emitters or the direction angles to the emitters is supported by a centralized fusion architecture. Fusion of the emitter angle track data is implemented with a decentralized architecture.

As shown in Figure 9.1, the first centralized fusion architecture combines the signal waveforms received at the antenna of a scanning surveillance radar, acting in a receive-only mode, with those from another passive receiver. The second passive receiver searches the same volume as the surveillance radar with a nonscanning, high-directivity multibeam antenna. The detection data obtained from the scanning and nonscanning sensors are used to calculate the unambiguous range to the emitters. This fusion approach allows the positions of the emitters to be updated at the same rate as data are obtained from the surveillance radar, making timely generation of the surveillance volume and emitter target location possible. One coherent processor is required for each beam in the multibeam antenna. A large communications bandwidth is also needed to transmit the radar signals to the multibeam passive receiver. The passive receiver is collocated with the coherent processors where the data fusion operations are performed.

In the second centralized fusion example, bearing angle-only data that describe the direction to the emitters are measured by multiple surveillance radars operating in a receive-only mode. The angle measurements are sent to a centralized location where they are associated to determine the unambiguous range to the emitters. Either elevation and azimuth angles or only azimuth angle measurements can be used as the input data to this fusion process. Ghost intersections, formed by intersecting radar beams for which targets do not exist, are eliminated by searching over all possible combinations of sensor angle data and selecting the most likely angle combinations for the target positions. The large number of searches needed to find the optimal measurement-target associations may require high processor throughput, which is a controlling factor in determining the feasibility of this fusion architecture when large numbers of emitters are present. The data association process is modeled using a maximum likelihood function. Two methods are discussed to solve the maximum likelihood problem. In the first method, the maximum likelihood process is transformed into its equivalent zero-one integer programming problem to find the optimal associations. In the second method, the computational requirements are reduced by applying a relaxation algorithm to solve the maximum likelihood problem. Although the relaxation algorithm produces somewhat suboptimal direction angle-emitter associations, in many cases they are within approximately 1 percent of the optimal associations.

The decentralized fusion architecture combines multiscan tracks that contain the direction angles to the airborne emitters that are output by the individual surveillance radars. The time history of the tracks aids in the calculation of the unambiguous range and eliminates the need for the large numbers of searches

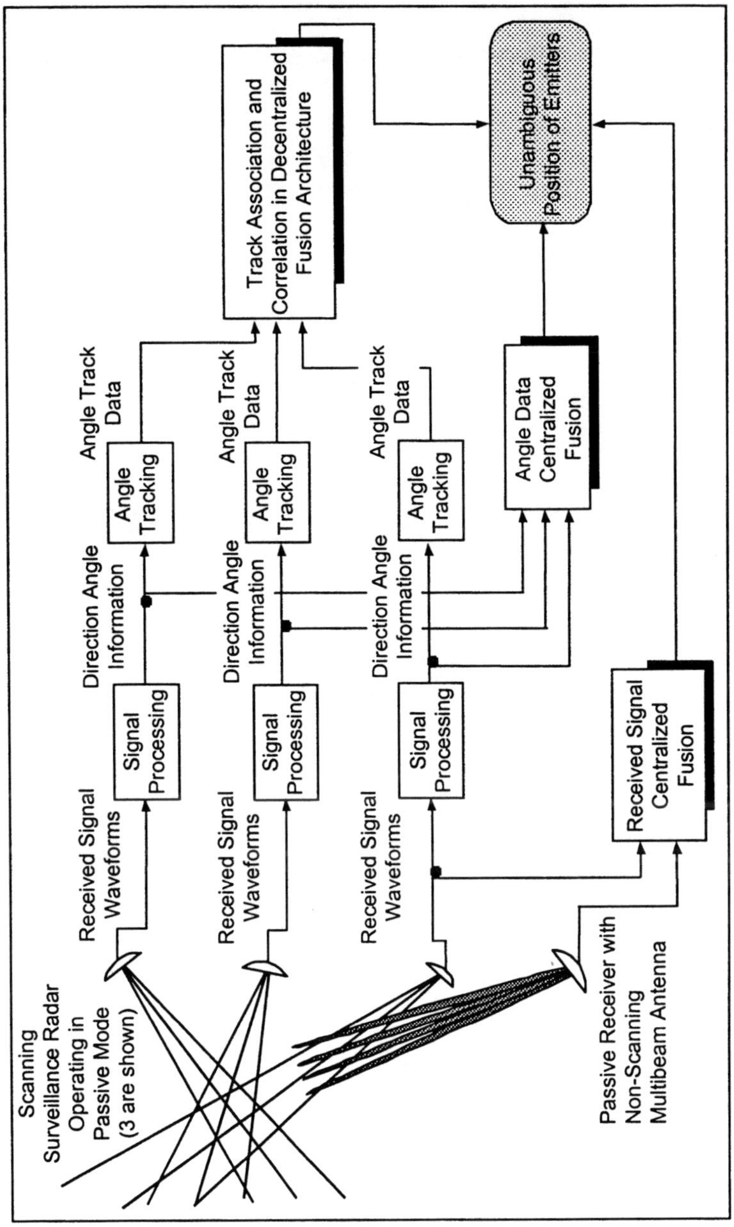

Figure 9.1: Passive sensor data association and fusion techniques for estimating location of emitters

that are required when deghosting is necessary. If angle tracks from only one passive sensor are available, it is still possible to estimate the range to the emitter if the tracking sensor is able to perform a maneuver. This latter case requires a six state Kalman filter as explained toward the end of the chapter.

All these techniques allow the unambiguous location of the emitters to be calculated. However, the impact on processing loads, communication bandwidths required for data transmission, and real-time performance differs. Advantages and disadvantages of each approach for processing passively acquired data are shown in Table 9.1 where a linkage is also made to the fusion architectures described in Chapter 3. Each of the techniques requires system-level trades as discussed in the appropriate sections below.

Table 9.1: Fusion techniques for associating passively acquired data to locate and track multiple targets

Fusion Level	Data Fusion Technique	Advantages	Disadvantages
• Received Signal (pixel-level fusion)	Coherent processing of data received from two types of sensors: a scanning surveillance radar and a passive receiver with a high-directivity multibeam antenna	• All available sensor information used • Unambiguous target location obtained • Data are processed in real time	• Large bandwidth communications channel required • Auxiliary sensor required • One coherent processor for each beam in the multibeam antenna required
• Angle Data (feature-level fusion)	Maximum likelihood or relaxation algorithm using direction angle measurements to the target	• 3D position of target obtained • Communication channel bandwidth reduced	• Ghosts created that have to be removed through increased data processing
• Target Track (decision-level or sensor-level fusion)	Combining of distributed target tracks obtained from each surveillance radar	• 3D position of target obtained • Communication channel bandwidth reduced even further	• Many tracks must be created, stored, and compared to eliminate false tracks

9.1 Received-signal fusion

The first centralized fusion architecture called received-signal fusion, combines the signal waveforms received by a scanning surveillance radar with those from a nonscanning (in this example) passive receiver that incorporates a multibeam antenna to search the volume of interest. The signals from these two sensors are transmitted to a central processor, where they are coherently processed to produce information that is used to locate the source of the signals.

The advantages of this architecture include unambiguous location of the emitter targets without creating ghosts that are characteristic of the angle data fusion architecture.[1] Ghosts occur when we believe there is a target present, but in truth there is no target present. Received-signal fusion requires transmission of large quantities of relatively high-frequency signal data to a centralized processor and, therefore, received-signal fusion places a large bandwidth requirement on the communications channel.

In the coherent processing technique illustrated in Figure 9.2, the scanning surveillance radar signals are combined with those from a multibeam antenna to compute the time delay and Doppler shift between the surveillance radar and multibeam antenna signals. These data, along with the instantaneous pointing direction of the surveillance radar, allow the estimation of the position and velocity of the emitters to be made, using triangulation techniques, for example.

The multibeam receive-only antenna is assumed to contain a sufficient number of beams to search the surveillance region of interest. This region is represented by the emitters indicated with "plus" symbols in Figure 9.2. The coherent processors operate jointly on the surveillance radar signal and the multibeam antenna signals to simultaneously check for the presence of emitters in all the regions formed by the intersecting beams. The ambiguity of declaring or not declaring the presence of an emitter in the observation space is minimized by the coherent processing. The multibeam antenna and the bank of coherent processors permit emitter positions to be calculated faster than is possible with angle data fusion. In fact, the emitter location information is available in real time as though the surveillance radar were making the range measurement by itself. In addition, coherent processing allows for simultaneous operation of the surveillance radar as an active sensor to detect targets in a nonjammed environment and also as a passive receiver to locate the emitters in a jammed environment.

9.1.1 Coherent processing technique

Knapp and Carter[2] and Bendat and Piersol[3] have suggested a method to reliably infer if one of the signals received by the multibeam antenna and the signal received by the surveillance radar emanate from a common source and are independent of a signal coming from another emitter. Their method treats the multibeam antenna and radar signals as random processes and calculates the dependence of the signal pairs using a cross-correlation statistic that is normalized by the energy contained in the two signals.

For our application, the Knapp and Carter cross-correlation statistic $\gamma(t)$ is given by

$$\gamma(t) = \frac{\left| \int_{t-T}^{t} x(t)\, y(t-\tau) \exp(-2\pi j v t) \mathrm{dt} \right|}{\left[\int_{t-T}^{t} |x(t)|^2 \mathrm{dt} \int_{t-T}^{t} |y(t-\tau)|^2 \mathrm{dt} \right]^2}, \tag{9-1}$$

where $x(t)$ and $y(t)$ represent the complex value of the two random processes (signals) over the immediate prior time interval T, which is equal to the signal processing time. The variables τ and v are estimates of the relative time delay and Doppler shift, respectively, between the signals received by the multibeam antenna and surveillance radar from the common emitter source.

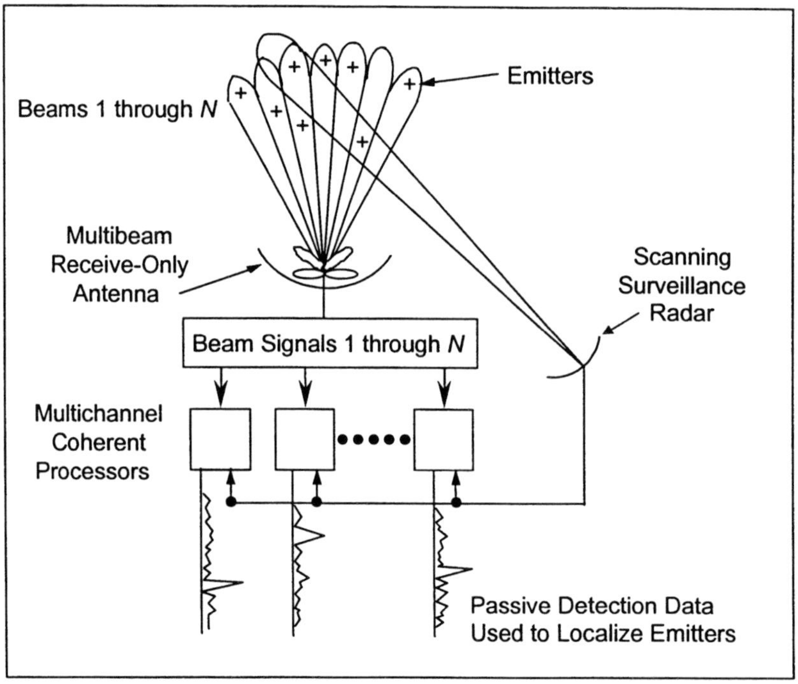

Figure 9.2: Coherent processing of passive signals

The $\gamma(t)$ statistic is particularly effective when the noise components in the signal are uncorrelated. In this case, Knapp and Carter show that the performance of a hypothesis test (deciding if an emitter is present or not) based on the cross-correlation statistic depends on (1) the signal-to-noise ratio (SNR) calculated from the power received at the multibeam antenna and the surveillance radar and (2) the time-bandwidth product formed by the product of the signal processing time T and the limiting bandwidth of the system. The limiting bandwidth is the smallest of the multibeam antenna receiver bandwidth, surveillance radar bandwidth, coherent processor bandwidth, or the communications channel bandwidth. In high-density emitter environments with relatively low SNRs, the cross-correlation statistic provides a high probability of correctly deciding if a signal is present for a given, but low value of the probability of falsely deciding that an emitter is present.

A typical result of the Knapp and Carter statistic for wideband coherent signals is shown in Figure 9.3. On the left are the real and imaginary parts of the signal received by the multibeam antenna. On the right are the corresponding signals received by the surveillance radar. The waveform at the bottom represents the output of the coherent processor. Estimates of the Doppler shift ν are plotted against estimates of the time delay τ. If the received signals come from the same emitter, then for some value of ν and τ there will be a large-amplitude sharp peak γ. If the peak is above a predetermined threshold, an emitter is declared present. The emitter's location is computed from the law of sines applied to the triangle formed by the baseline distance between the surveillance radar and the multibeam receiver, and the azimuth direction angles to the emitter as measured by the surveillance radar and multibeam receiver. The trigonometry for the calculation is shown in Figure 9.4. Since the radar is rotating, the relative time delay τ gives a correction for the azimuth direction angle of the radar in the law of sines range calculation. The elevation of the emitter is also known from the sensor data.

9.1.2 System design issues

The primary subsystems in the received-signal fusion architecture are the surveillance radar, the multibeam antenna including its beamforming network, the communication link between the surveillance radar and the coherent processors, and the coherent processors themselves. Table 9.2 lists the major issues that influence the design of the coherent-receiver fusion architecture.

The system's complexity and performance are determined by the relationships between the design parameters. Complexity is primarily affected by the throughput requirements for the coherent processor, the design of the passive multibeam antenna, and the bandwidth and jam resistance of the surveillance radar-to-processor communication channel. Throughput requirements for the coherent processor depend on the number of beams, the number of time-delay and Doppler-shift cells that must be searched for a maximum in the cross-correlation signal, and the processing gain required for the hypothesis test that determines whether the signals emanate from a common source. The number of beams is dependent on the resolution of the multibeam antenna and its angular field of view. Processing gain depends on SNR, which in turn depends on the spatial and amplitude distributions of the emitters in relation to the angular resolution of the radar and multibeam antenna.

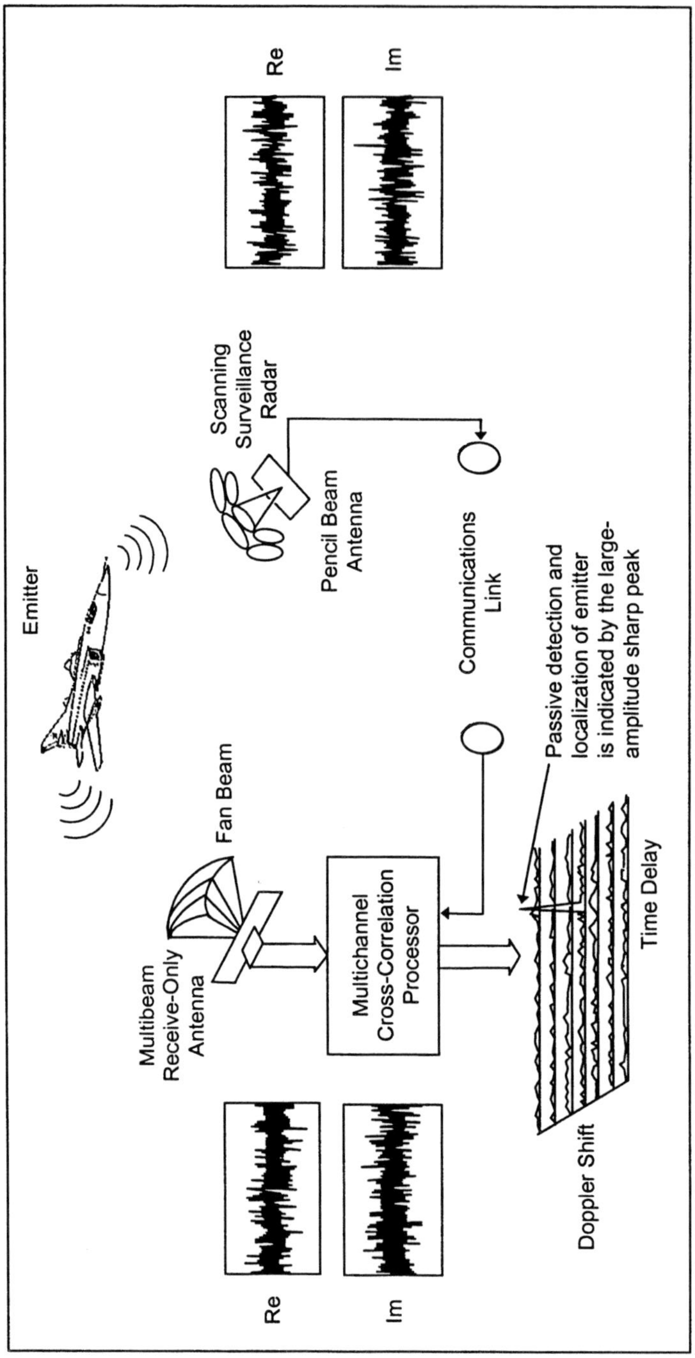

Figure 9.3: Cross-correlation processing of the received passive signals

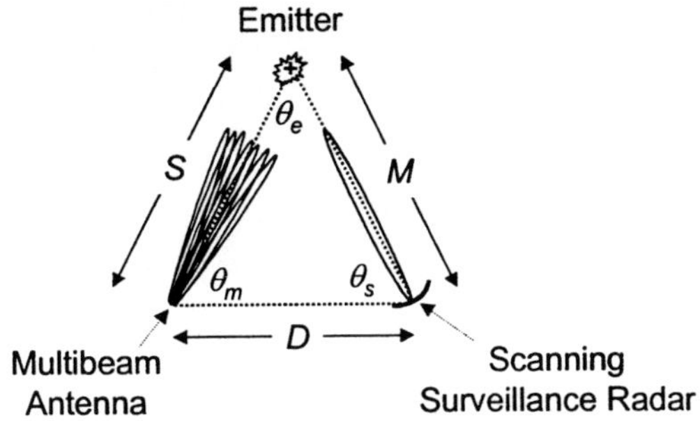

From law of sines:

- $D/\sin\theta_e = M/\sin\theta_m = S/\sin\theta_s$
- D, θ_m, θ_s are known from measurements
- $\theta_e = 180 - \theta_m - \theta_s$
- Distance S from multibeam antenna to emitter is

  $S = D \sin\theta_s / \sin\theta_e$
- Distance M from surveillance radar to emitter is

  $M = D \sin\theta_m / \sin\theta_e$

Figure 9.4: Law of sines calculation of emitter location

Table 9.2: Major issues influencing the design of the coherent receiver fusion architecture

Issue	Design Impact
• Spatial and amplitude distribution of emitters	• Angular resolution • Baseline separation • Processing gain
• Emitter velocity	• Number of Doppler cells
• Coherence of transmission media as it affects emitter signals	• Processing gain
• Angular resolution of surveillance radar and multibeam antenna	• Number of time-delay cells • Signal-to-noise ratio
• Baseline separation between surveillance radar and multibeam antenna	• Communications requirements (amplifiers, repeaters, noise figure, etc.) • Number of time-delay cells
• Processing gain	• Throughput of coherent processors
• Simultaneous operation of radar and multibeam antenna	• Signal-to-noise ratio
• Directivity of radar and multibeam antenna	• Number of coherent processors • Sensitivity of multibeam antenna receiver
• Resistance to jamming of baseline communications channel	• Communications techniques (spread spectrum, time-division multiple access [TDMA], etc.)

The range of time delays that must be searched in the coherent processor depends directly on the angular resolutions of the radar and multibeam antenna.[4] The number of time-delay cells is proportional to the total amount of delay normalized by the signal observation interval *T*. The upper bound of the observation interval is given by the radar angular resolution divided by its angular search rate.

The range of Doppler shift that must be searched to locate emitters depends on the angular field of view of the system.[4] The number of Doppler-shift cells in the coherent processor is proportional to the total amount of Doppler shift normalized by the signal bandwidth. For broadband emitters, the upper bound to the signal bandwidth is given by the bandwidth of the radar transfer function. Within the constraints imposed by the radar, it is feasible to independently choose various values for observation interval and signal bandwidth, such that their product equals the required time-bandwidth product for the coherent processing. The computations associated with the coherent processing are minimized when the observation interval and signal bandwidth are optimized through trades among observation interval, number of time-delay cells, bandwidth, and number of Doppler-shift cells.

Surface and volume clutter will adversely affect the SNR at both the surveillance radar and multibeam antenna. The quantitative effects depend on the effective radiated power and directivity of the radar, the directivity of the multibeam antenna, and the amplitude distribution characteristics of the clutter in the radar's field of view. Coherent processor performance is affected by the amplitude and phase components of the signal at the input to the coherent processor. The signal bandwidth, in turn, depends on the transmission medium's temporal and spatial coherence statistics, the nonlinearities of the radar and multibeam antenna response functions, and the amplitude and phase transfer functions of the baseline communications channel.

The distance between the radar and multibeam antenna affects the performance of the fusion system in four significant ways: (1) radar-to-multibeam antenna communication requirements including jammer resistance and signal amplification and filtering, (2) range of time delays that must be searched by the coherent processor, (3) mutual surveillance volume given by the intersection of the radar and multibeam antenna fields of view, and (4) accuracy with which the emitters are located. Typical separation distances between the radar and multibeam antenna are 50 to 100 nautical miles (93 to 185 km). In addition, the topography along the radar-to-multibeam antenna baseline influences the applicability of a ground-to-ground microwave communications link.

9.2 Angle data fusion

In the second centralized fusion architecture, referred to as angle data fusion, multiple surveillance radars (operating in a receive-only mode) are used to measure the elevation and azimuth angles that describe the direction to the emitters. These data are fused in a central processor to find the number of real emitters and estimate the unambiguous range to each. Associating multisensor data at a given time instant, as required in this fusion architecture, is analogous to associating data from the successive scans of a single sensor to form tracks.[5]

The main design elements of the passive surveillance radar system are the antenna, the detection and data association processes, and the communication link between the radars and the central processing unit. IRST sensors can also passively track these targets. When they are used, the sensor separation can be reduced to between 10 and 15 nautical miles (19 to 28 km) because of the IRST's superior angle measurement accuracy as compared to microwave radars.

9.2.1 Solution space for emitter locations

If there are M radar receivers and N emitters in the field of view of the radars, then associated with each emitter is an M-tuple of radar direction angle measurements that uniquely determines the position of the emitter. When the number of direction angle measurements made by each radar is equal to N, there are as many as N^M unique direction M-tuples, or potential emitter locations, to sort through since the true position of the emitters is unknown.

Not all M-tuple combinations represent real locations for the emitters. For example, there are M-tuples that will place multiple emitters at the same direction angle and, thereby, invalidate the number of independent measurements known to be made by a particular radar. This is illustrated in Figure 9.5.

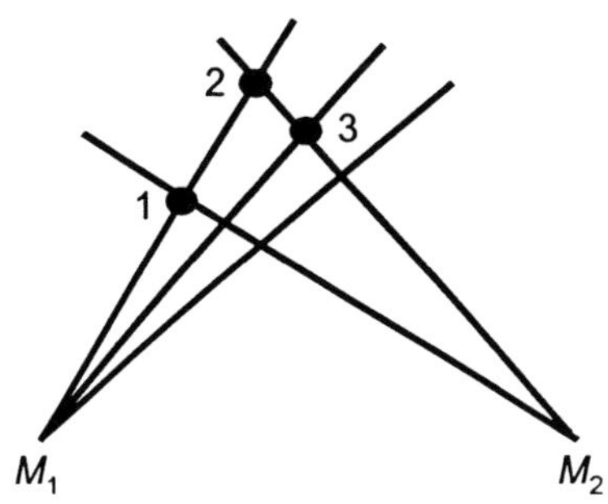

Figure 9.5: Unacceptable emitter locations

Three emitter locations are known to have been detected by the radar on the left as represented by the three direction angle measurements emanating from M_1.

Two emitter locations are known to have been detected by the radar on the right as represented by the two direction angle measurements emanating from M_2. The detection of only two emitters by the radar on the right can occur when two of the three emitters lie on the same direction angle or the radar's resolution is inadequate to resolve the emitters. In Figure 9.5, Emitters 1 and 2 are placed on the left-most direction angle and Emitter 3 on the middle direction angle measured by Radar 1, leaving no emitters on the right-most direction angle. This combination represents a fallacious solution that must be excluded since the premise of three direction angle measurements by Radar 1 is not represented. The false positions are eliminated by constraining the solution to contain the same number of emitter direction angle measurements as corresponds to the radar data and to use each angle measurement only once. Since there are N emitters, there are only N true positions to be identified. Thus there are N^M - N ambiguous M-tuple locations to be eliminated, because these represent locations for which there are no emitters.

When the number of direction angle measurements from each of the radars is not equal, the number of potential locations for the emitters must be found in another manner. The procedure for this case is illustrated by the example of Figure 9.6.

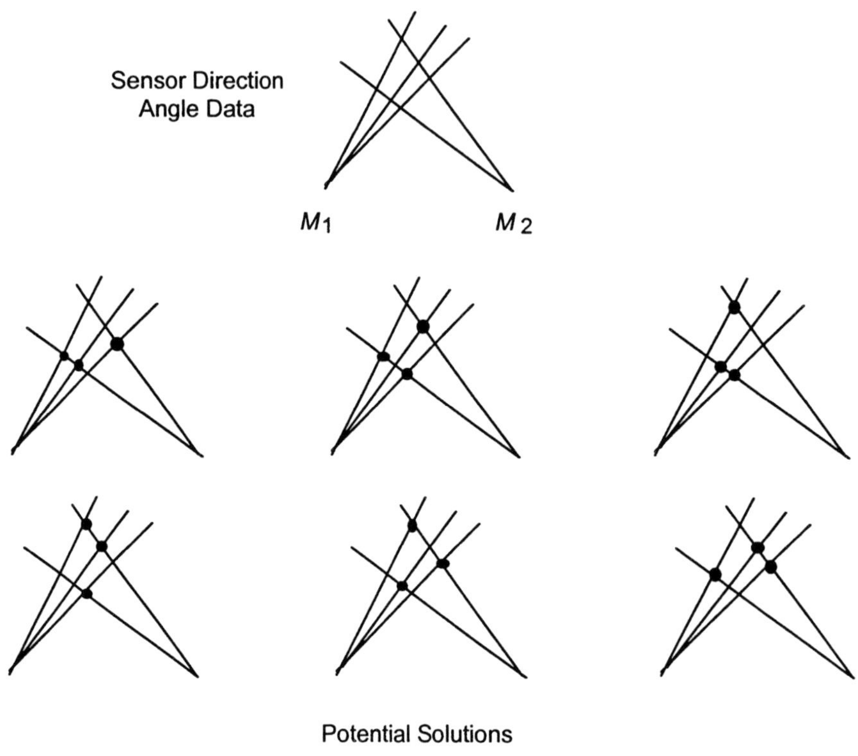

Figure 9.6: Ambiguities in passive localization of three emitter sources with two receivers

Six potential locations for three emitters jamming two radar receivers are illustrated. However, only one set of intersections formed by the direction angle measurements corresponds to the real location of the emitters. The upper part of the figure shows that the first radar measures angle data from all three emitters as indicated by the three lines whose direction angles originate at point M_1. The number of angle measurements is denoted by $N_1 = 3$. The second radar, due to its poorer resolution or the alignment of the emitters or both, measures angle data from only two emitters as shown by the two direction angles that originate at point M_2. Here the number of angle measurements is denoted by $N_2 = 2$. The total number of intersections is equal to $N_1 \times N_2 = 6$. The six potential solutions that result are illustrated in the lower portion of the figure. The problem is to identify the solution that gives the best estimate for the location of the three emitters.

Figure 9.7 illustrates the ambiguities that arise for a generalization to N emitters and three radars. The upper portion of the figure shows the number of angle measurements made by each radar, namely, N_1, N_2, and N_3. The lower left shows the intersection of the three radar beams with the N emitters. The total number of intersections is given by $N_1 \times N_2 \times N_3$.

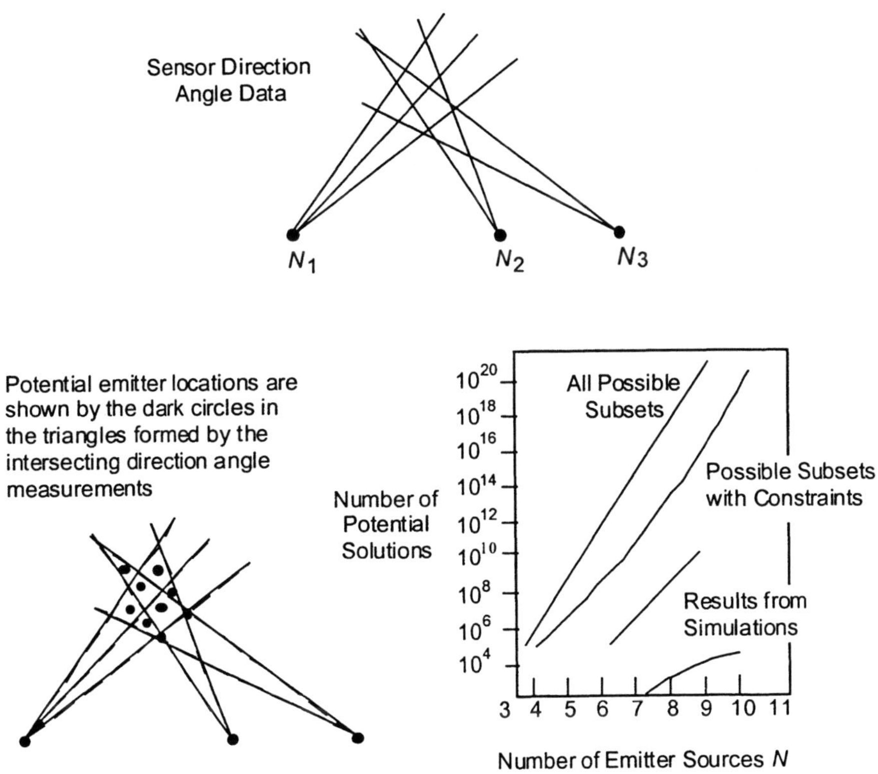

Figure 9.7: Ambiguities in passive localization of *N* emitter sources with three receivers

The graph in the lower right of the figure contains four curves. The upper curve, labeled "All Possible Subsets," represents the N^M unique solutions that correspond to the direction angle measurements made by each of the three radars. The curve labeled "Possible Subsets with Constraints" represents the number of unique solutions assuming that an angle measurement is used only once to locate an emitter. The bottom two curves are from simulations that use prefiltering to remove unlikely intersections. The prefilter examines the intersection space formed by the radar direction angle measurements and eliminates those having intersection areas greater than some preset value. Intersections located behind any of the radars are also eliminated. Clearly the prefiltering reduces the number of potential solutions. Using an efficient search algorithm with the prefilter (efficient in terms of the number of iterations required to reach an optimal or near optimal solution), such as the set partitioning and relaxation algorithms discussed in the next section, reduces the number of potential solutions even further as shown in the bottom curve. However, the number of potential solutions remains large (of the order of 10^4), even for the modest numbers of emitter sources shown.

If both azimuth and elevation angle data are measured by the radars, then the search algorithm is simplified considerably. In this case, a two-dimensional assignment algorithm can be used and the requirement for a three-radar system is reduced to a two-radar system.

Algorithm performance and throughput are determined by the numbers of densely positioned emitter sources and the radar resolution. In these environments, algorithms must (1) make consistent assignments of radar angle measurements to emitter positions while minimizing ghost and missed emitter positions, (2) fuse direction angle information from three or more radars that possibly have poor resolution in an environment where multipath may exist, and (3) be efficient for real-time or near real-time applications in their search of the number of potential solutions and the assigning of M-tuples to emitters.

The first data association technique discussed for the three-radar system uses zero-one integer programming to find the optimal solution by efficiently conducting a maximum likelihood search among the potential M-tuples. The azimuth direction measurements obtained from the radars are assigned to the N emitter locations with the constraint that each angle measurement be used only once in determining the locations of the emitters. Prefiltering is employed to reduce the direction angle-emitter association (M-tuple) space.

The second technique uses a relaxation algorithm to speed the data association process that leads to the formation of M-tuples. The relaxation algorithm produces suboptimal solutions, although simulations have shown these angle measurement associations to be within 1 percent of the optimal.

9.2.2 Zero-one integer programming algorithm development

Consider a planar region where N emitters or targets are described by their Cartesian position (x, y). Assume that the targets lie in the surveillance region of the radars and are detected by three noncollocated radar sensors (having known positions and alignment) that measure only azimuth angle Θ from the emitter relative to north. The statistical errors associated with each radar's directional measurement are assumed to be Gaussian distributed with zero mean and known variances. In addition, spurious directional measurement data, produced by phenomena such as multipath, are present and are uniformly distributed over the field of view of each sensor. We call an emitter location estimable if all three radars detect the direction angle (in this case the azimuth angle) to the emitter. We shall calculate the positions for only the estimable emitters, but not for those that are unresolved by the radars.

The solution involves partitioning the angle measurements into two sets, one consisting of solutions corresponding to the estimable emitter positions and the second corresponding to spurious measurements. Spurious data are produced by multipath from azimuth angle measurements and the N^3 minus N 3-tuples that represent ambiguous positions generated by the incorrect association of azimuth angles.

Partitioning by the maximum likelihood function selects the highest probability locations for the emitters. The maximum likelihood function L is the joint probability density function corresponding to the emitter locations. It is given by[6]

$$L = \prod_{\gamma \in \Gamma} \frac{1}{(2\pi)^{3/2}} |\Sigma|^{-1/2} [\exp(-1/2\,\Theta_\gamma^{\mathrm{T}}\, \Sigma^{-1}\, \Theta_\gamma)](1/\Phi_1)^{m_1 - N} (1/\Phi_2)^{m_2 - N} (1/\Phi_3)^{m_3 - N} \tag{9-2}$$

where

Γ = set of all possible 3-tuples that represents the real and ambiguous emitter locations,

γ = 3-tuple of radar angle data that is believed to correspond to a particular emitter,

$\Sigma = \mathrm{diag}\,[\sigma_1^2, \sigma_2^2, \sigma_3^2]$,

σ_r^2 = variance of the angle measurements associated with the rth radar (r = 1, 2, and 3 in this example),

Φ_r = field of view of the rth radar, $0 \le \Phi_r \le 2\pi$,

m_r = number of direction measurements associated with Radar r for one revolution of the radar,

N = number of emitters,

Θ_γ = particular 3-tuple vector of direction angle measurements from Radars 1, 2, and 3,

= $[\Theta_{1i}, \Theta_{2j}, \Theta_{3k}]^{\text{T}}$ where $i, j,$ and k refer to a particular direction angle measurement from Radars 1, 2, and 3, respectively, and

T = transpose operation.

To facilitate the search over all possible 3-tuples, the maximum likelihood problem is replaced with its equivalent zero-one integer programming problem. The zero represents nonassignment of direction angle measurements to a 3-tuple, while the one represents assignment of direction angle measurements to a 3-tuple, with one direction angle being assigned from each radar.

The maximization of the likelihood function is equivalent to minimizing a cost function given by the negative of the natural logarithm of the likelihood function shown in Eq. (9-2). The use of the cost function and a set of constraints allows the original problem to be solved using the zero-one integer programming algorithm.

When the fields of view of the three radars are equal such that $\Phi_1 = \Phi_2 = \Phi_3 = \Phi$, the cost function C can be written in the form

$$C = (m_1 + m_2 + m_3)\ln\Phi + \sum_{i=1}^{m_1}\sum_{j=1}^{m_2}\sum_{k=1}^{m_3}(C_{ijk} - 3\ln\Phi)\delta_{ijk} \tag{9-3}$$

where

$$C_{ijk} = \Theta_\gamma^{\text{T}}\,\Sigma^{-1}\,\Theta_\gamma \tag{9-4}$$

subject to

$$\sum_i\sum_j \delta_{ijk} \le 1 \text{ for all } k, \tag{9-5}$$

$$\sum_i\sum_k \delta_{ijk} \le 1 \text{ for all } j, \tag{9-6}$$

and

$$\sum_{j}\sum_{k}\delta_{ijk} \leq 1 \text{ for all } i, \tag{9-7}$$

where

$$\delta_{ijk} = 1 \tag{9-8a}$$

when the ith direction angle from Radar 1, the jth from Radar 2, and the kth from Radar 3 are selected, and

$$\delta_{ijk} = 0 \tag{9-8b}$$

when these direction angles are not selected.

The constraint is to use an angle measurement from a radar only once in forming the 3-tuples. This constraint may cause an emitter location to be missed when the radar resolution is not adequate to provide one measurement for each emitter, or when the emitters are aligned such that some are blocked from the view of the radars. These drawbacks will, over time, resolve themselves due to emitter motion and the geometry of the search situation.

Throughput requirements can be reduced by eliminating solutions that make the term (C_{ijk} - 3 lnΦ) positive, such as by preassigning $\delta_{ijk} = 0$ for these solutions, because this always decreases the value of the cost function. With the above constraint and the elimination of positive cost function solutions, the zero-one integer programming problem is converted into the standard set-packing problem formulation,[7] solved by using any set-partitioning algorithm such as those described by Pierce and Lasky[8] and Garfinkel and Nemhauser.[9] Further simplification is made by eliminating still other variables, such as those representing small costs, even though they are negative. This produces a suboptimal 3-tuple, but considerably reduces the number of searches required to solve the zero-one integer programming problem. Since three radars are used in this example, the integer programming is solved with a three-dimensional assignment algorithm as described by Frieze and Yadegar.[10]

Figures 9.8 through 9.10 show the results of applying the above techniques to a scenario containing 10 emitters and 3 radars. The emitters, referred to as the "True Targets," were randomly placed along a racetrack configuration as shown by the dark squares in Figure 9.8. The racetrack was approximately 60 nautical miles (111 km) north of the radars located in (x, y) coordinates at (-50, 0), (0, 0), and (50, 0) nautical miles (50 nautical miles equals 93 km) as depicted by the "spade" symbols along the x-axis. Radar resolution was modeled as 2 deg. The standard deviation of the radar angle measurements was assumed to be 0.5 deg.

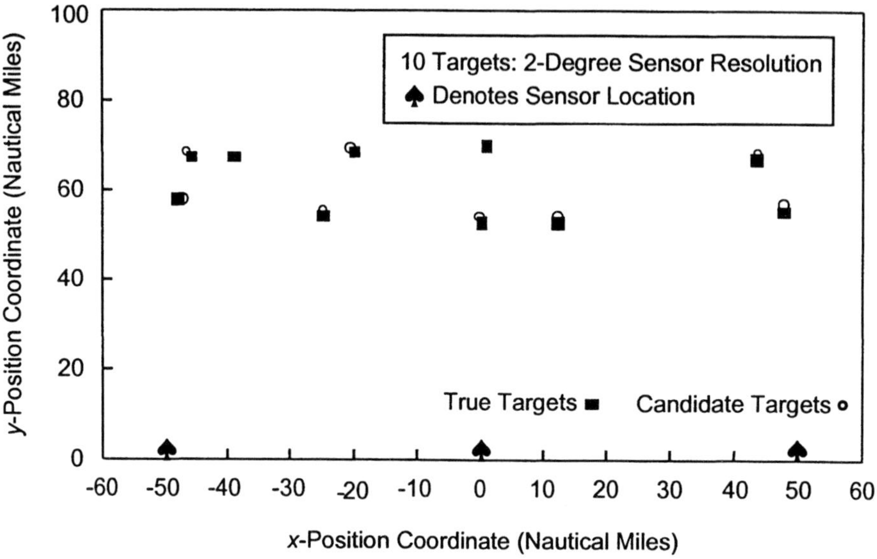

Figure 9.8: Passive localization of 10 emitters using zero-one integer programming

Figure 9.9 shows all the possible subsets of candidate target positions, represented by open circles before prefiltering and the other constraints were applied. In Figure 9.10, the number of possible target positions processed by the zero-one integer programming algorithm has been reduced by prefiltering. The final result of applying the cost constraints and the zero-one integer programming is shown in Figure 9.8 by the open circles. Positions of 8 of the 10 true emitter targets were correctly identified. The two targets that were not located lie within 2 deg of each other and, therefore, could not be detected with the radar resolution limit of 2 deg used in this example. Since the targets are moving, however, this system would be able to resolve all the emitter targets as their separation increased with time to beyond 2 deg.

High-speed computers reduce the computation time required by the zero-one integer programming approach. A calculation with a Macintosh IIsi incorporating a Motorola 68030 20-MHz processor and coprocessor provided solutions to the 10-target problem using 2 to 3 s of central processor unit (CPU) time per radar scan.[11] Another host using a Sky Computers Skybolt VME board containing one i860 processor (80 MFLOPS) reduced the CPU time to less than 0.2 s with 10 targets and less than 1.3 s with 20 targets.

These CPU usage times are per run averages based on 10 runs. Faster executing microprocessors will reduce the time even further. Since long-range surveillance radars have scan rates of about 10 s, it is feasible to implement these algorithms in near real time with a restricted number of targets. However, as the number of

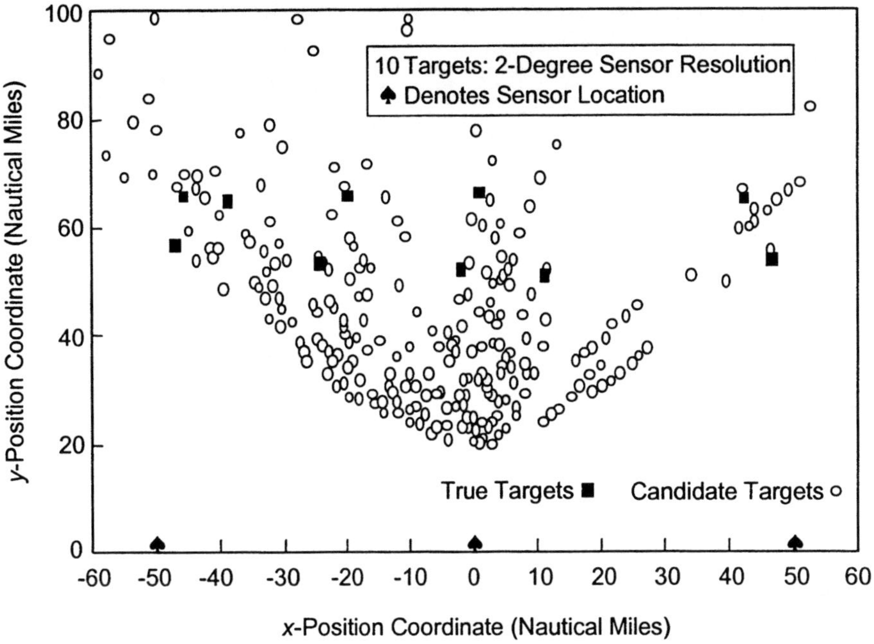

Figure 9.9: All subsets of possible emitter positions before prefiltering and cost constraints are applied

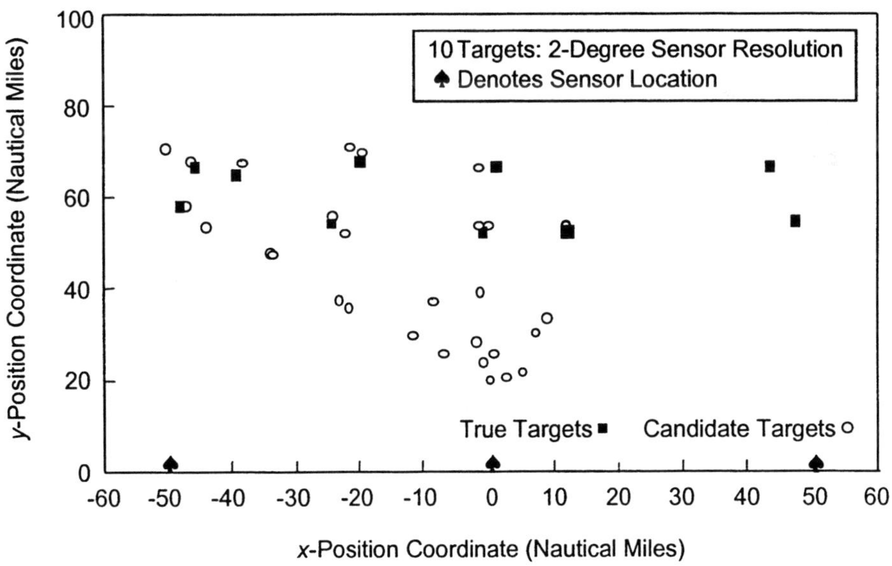

Figure 9.10: Potential emitter positions that remain after prefiltering for input to zero-one integer programming algorithm

potential targets increases, real-time execution of the zero-one integer programming technique becomes more difficult. This is due to the exponential increase in the complexity of the optimal algorithm with the number of measurements made by each sensor, since the algorithm has nonpolynomial computational time and memory requirements. Suboptimal algorithms such as the relaxation algorithm are, therefore, of considerable importance as they require smaller computational times.

9.2.3 Relaxation algorithm development

A recently developed method of speeding the search through potential solutions uses a Lagrangian relaxation algorithm.[12] With this approach, near optimal solutions (producing sensor data association *M*-tuples within approximately 1 percent of the optimal) can be obtained for reasonable computing times with moderate numbers of emitters (of the order of 20). In the development by Pattipati et al., unconstrained Lagrange multipliers are used to reduce the dimensionality of the three-dimensional data assignment problem to a series of two-dimensional assignment problems. A heuristic strategy that recomputes the data association at every iteration of the solution minimizes the cost of the suboptimal solution as compared with the optimal. A desirable feature of this method is the estimate it produces of the error between the feasible suboptimal solution and the global optimal solution. The error may be used to control the number of iterations.

Algorithm run time is a function of the sparsity of the search volume and the number of reports from each sensor. Sparsity is defined as the ratio of the average number of potential direction angle measurement-emitter associations in the three-dimensional assignment problem to the number of angle measurement-emitter associations required for a fully connected graph. The graph represents the *M*-tuple associations of the sensor measurements with emitters. For example, if a graph has 10 nodes (where a node is the number of reports per sensor), there are $10^3 = 1000$ angle measurement-emitter associations with a three-sensor system. If there are *M* angle measurement-emitter associations instead, the sparsity of the graph *S* is

$$S = \frac{M}{1000}. \tag{9-9}$$

Therefore, a larger value of *S* implies a greater graph density.

Data in Table 9.3 (from Pattipati et al.) show the speedup provided by the relaxation algorithm over a branch-and-bound algorithm[13] averaged over 20 runs. (The algorithm described by Pierce and Lasky[8] also provides improved results over the branch-and-bound, but not as much as the relaxation algorithm.)

Table 9.3: Speedup of relaxation algorithm over a branch-and-bound algorithm (averaged over 20 runs)
(K.R. Pattipati et al., *IEEE Trans. Auto. Cont.*, 37(2), 198-213 [Feb. 1992])

Number of Reports From Each Sensor	Sparsity = 0.05	Sparsity = 0.1	Sparsity = 0.25	Sparsity = 0.5	Sparsity = 1.0
5	1.2 (0.006)*	1.6 (0.01)	1.7 (0.04)	2.2 (0.2)	16.6 (0.4)
10	3.0 (0.02)	3.8 (0.06)	30.3 (0.6)	3844.0 (1.4)	† (2.3)
15	4.8 (0.16)	26.4 (0.27)	1030.4 (2.1)	† (3.6)	† (5.8)
20	18.5 (0.2)	656.1 (0.44)	† (2.3)	† (7.5)	† (12.1)

* The numbers in parentheses denote the average time, in seconds, required by the relaxation algorithm on a Sun 386i workstation.

† Denotes that memory and computational resources required by the branch-and-bound algorithm exceeded the capacity of the Sun 386i (5 MIPS, 0.64 MFLOPS, 12 Mb RAM) workstation.

Branch-and-bound algorithms perform a structured search for a solution. They are based on enumerating only a small number of the possible solutions, because the remaining solutions are eliminated through the application of bounds. The branch-and-bound algorithm involves two operations: branching, i.e., dividing possible solutions into subsets, and bounding, i.e., eliminating those subsets that are known not to contain solutions. The basic branch-and-bound technique is a recursive application of these two operations.[14,15]

The branch-and-bound algorithm was found impractical for graphs containing 500 or more angle measurement-emitter associations. Hence, the speedup was not computed for these cases. The average run time of the relaxation algorithm on a Sun 386i is shown in parentheses. For these particular examples, the average solution to the angle measurement-emitter association problem was within 3.4 percent of optimal. As the sparsity decreases, the percent of suboptimality also decreases. In another example cited by Pattipati, the worst case suboptimal solution was within 1.2 percent of the optimal when the sparsity was 0.25 and the number of reports per sensor was 10.

Although the relaxation algorithm provides fast execution, there is no guarantee that an optimal or near optimal solution with respect to cost gives the correct association of angle measurements to emitters. In fact, as the sensor resolution deteriorates, it becomes more difficult to distinguish the true emitters from ghosts.

9.3 Decentralized fusion architecture

The decentralized fusion architecture finds the range to the emitters from the direction angle tracks computed by receive-only sensors located at multiple sites. These tracks are formed from the autonomous passive azimuth and elevation angle data acquired at each site. The tracks established by all the sensors are transmitted to a fusion center where they are associated using a metric. Examples of metrics that have been applied are the distance of closest approach of the angle tracks and the hinge angle between a reference plane and the plane formed by the emitters and the sensors.[16] The range information is calculated from trigonometric relations that incorporate the measured direction angles and the known distances between the sensors.

A number of Kalman filter implementations may be applied to estimate the angle tracks. In the first approach, each sensor contains a multistate Kalman filter to track azimuth angles and another to track elevation angles. The number of states is dependent on the dynamics of the emitter. In another approach, the azimuth and elevation angle processing are combined in one filter, although the filters and fusion algorithms are generally more complicated in this case. In either implementation, the azimuth and elevation angle Kalman filters provide linearity between the predicted states and the measurement space because the input data (viz., azimuth and elevation angles) represent one of the states that is desired and present in the output data.

Data analysis begins at each sensor site by initiating the tracks and then performing a sufficient number of subsequent associations of new angle measurement data with the tracks to establish track confidence. The confidence is obtained through scan-to-scan association techniques such as requiring n associations out of m scans. An optimal association algorithm, such as an auction algorithm, can be used to pair emitters[17] seen on scan n to emitters observed on the following scan $n+1$. The set of emitters on scan $n+1$ that are potential matches are those within the maximum relative distance an emitter is expected to move during the time between the scans. A utility function is calculated from the distance between the emitters on the two scans. The auction algorithm globally maximizes the utility function by assigning optimal emitter pairings on each scan.

The performance of the auction algorithm is shown in Figure 9.11. The average association error decreases as the sensor resolution increases and the interscan time decreases. The angle tracks produced at an individual sensor site are not sufficient by themselves to determine the localized position of the emitters. It is necessary to transmit the local angle track files to the fusion center, where redundant tracks are combined and the range to the emitters is computed and stored in a global track file.

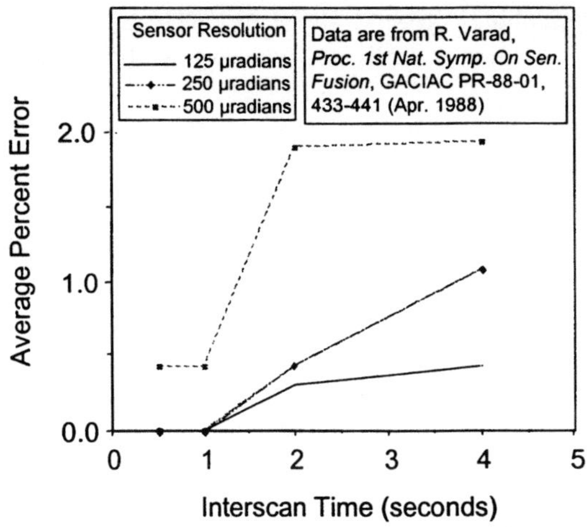

Figure 9.11: Average scan-to-scan association error of auction algorithm over 15 scans
(S.E. Kolitz, *Proc. SPIE* 1481, 329-340 [1991])

9.3.1 Local optimization of direction angle track association

The simplest decentralized architecture fuses emitter tracks by first estimating the time histories of the angle tracks produced by each sensor and then pairing them using a metric that measures the distance between the tracks. Tracks are associated when the metric is less than a preselected threshold. This technique does not globally optimize the track association among the sensors because the track pairings are not stored or recomputed after all tracks and data have been analyzed. Local track optimization was used in early air defense systems to track the objects detected by the sensors.

To locally optimize track association, the first track produced by Sensor 1 is selected and compared with the first track produced by Sensor 2. A metric such as the chi-squared (χ^2) value of the distance at the point of closest approach of the direction angle tracks (which is equal to the Mahalanobis distance) is calculated for each pairing. If the value exceeds a threshold, the pairing is discarded since the threshold exceedance indicates that the particular pairing will not produce the desired probability that the two tracks are from the same emitter. The track from Sensor 1 is then compared with the next track from Sensor 2. The process continues until the χ^2 value is less than the threshold. At this point, the angle tracks are combined as they are believed to represent the same emitter. *Paired tracks from Sensors 1 and 2 are removed from the lists of available tracks and the next track from Sensor 1 is selected for association with a track from Sensor 2 that is still unpaired.*

The process continues until all tracks from Sensor 1 are associated or until the list is exhausted. Unpaired tracks are retained for later association with the unpaired tracks from the other sensors. If a third set of angle tracks is available from a third sensor, they are associated with the fused tracks from the first two sensors by repeating the above process. Again unpaired tracks are retained. After all the sets of available angle tracks have been through the association process, the unpaired tracks from one sensor are compared with unpaired tracks from sensors other than the one used in its initial parings. If the χ^2 value of the distance at the point of closest approach of the tracks is less than the threshold, the tracks are paired and removed from the unpaired list. Unpaired tracks arise because one sensor may not detect the target during the time another sensor measured a track. The technique just described is analogous to a first-in, first-out approach with respect to the selection of pairings for the tracks from Sensor 1.

9.3.2 Global optimization of direction angle track association

Two methods of globally optimizing the association of the direction angles measured by the sensors will be discussed. The first applies a metric based on the closest approach distance of the direction angle tracks. The second uses the hinge angle. Global optimization is achieved at the cost of some increased computational load as compared with the local optimization method.

Closest approach distance metric

To globally optimize the track association with the closest approach distance metric, a more complex algorithm is needed such as the Munkres algorithm,[18] its extension by Bourgeois and Lassalle,[19] or the faster executing JVC[20] algorithm. The advantage of these algorithms comes from postponing the decision to associate tracks from the various sensors until all possible pairings are evaluated. In this way track pairings are globally optimized over all possible combinations. The process starts as before by selecting the first track from Sensor 1 and comparing it with the first track from Sensor 2. If the χ^2 value for the closest approach distance of the track direction angles exceeds the threshold, the pairing is discarded and the track from Sensor 1 is compared with the next track from Sensor 2. The process continues until χ^2 is less than the threshold. At this point, the angle tracks are combined as they are believed to represent the same emitter and the value of χ^2 is entered into a table of track pairings. However, the paired track from Sensor 2 is not removed from the list as before. *Also, the process of pairing the first track from Sensor 1 with those of Sensor 2 continues until all the tracks available from Sensor 2 are exhausted.* Whenever χ^2 is less than the threshold value, the χ^2 value corresponding to the pairing is entered into the table of pairings. This approach can therefore identify more than one set of potential track pairings for each track from Sensor 1.

Then the next track from Sensor 1 is selected for association and compared with the tracks from Sensor 2 beginning with the first track from the Sensor 2 track list. *Tracks from Sensor 2 used previously are made available to be used again in this algorithm.* If the χ^2 value exceeds the threshold, the pairing is discarded, and the track from Sensor 1 is compared with succeeding tracks from Sensor 2. When the χ^2 value is less than the threshold, that value is entered into the table of track pairings.

The process continues until all tracks from Sensor 1 are associated or until the lists are exhausted. Unpaired tracks are retained for later association with the unpaired tracks from the other sensors. If a third set of angle tracks is available from a third sensor, they are associated with the fused tracks from the first two sensors by repeating the above process.

Global optimization of the track pairings occurs by using the Munkres or the faster executing JVC algorithm to examine the χ^2 values in the table that have been produced from all the possible pairings. Up to now, the χ^2 value only guarantees a probability of correct track association if the angle tracks are used more than once. The Munkres and JVC algorithms reallocate the pairings in a manner that minimizes the sum of the χ^2 values over all the sensor angle tracks and, in this process, the algorithms use each sensor's angle tracks only once.

The acceptance threshold for the value of χ^2 is related to the number of degrees of freedom n_f which, in turn, is equal to the sum of the number of angle track measurements that are paired in the Sensor 1 tracks and the Sensor 2 tracks. Given the desired probability for correct track association, the χ^2 threshold corresponding to n_f is determined from a table of χ^2 values.

Hinge angle metric

The hinge angle metric allows immediate association of detections to determine the emitter position and the initiation of track files on successive scans. Calculations are reduced, as compared to the Munkres algorithm, by using a relatively simple geometrical relationship between the emitters and the sensors that permits association of detections by one sensor with detections by another. The metric allows ordered sequences of emitters to be established at each sensor, where the sequences possess a one-to-one correspondence for association.

Processing of information from as few as two sensors permits computation of the hinge angle and the range to the emitters. However, each sensor is required to have an attitude reference system that can be periodically updated by a star tracker or by the Global Positioning System, for example. It is also assumed that the sensors have adequate resolution to resolve the emitters and to view them simultaneously. Occasionally, multiple emitters may lie in one plane and may not be resolvable by all the sensors. However, the emitters will probably be resolved during subsequent scans. The use of three sensors mitigates this problem.

The hinge angle procedure defines a unique emitter target plane based on the line-of-sight (LOS) vector from a sensor to a particular emitter and the line joining the two sensors as shown in Figure 9.12. Each emitter target plane contains the common LOS vector between the two sensors and each can be generated by a nominal reference plane rotated about the LOS vector between the two sensors. The angle between the emitter target plane and the reference plane is called the hinge angle. The reference plane is defined by Varad[16] to contain the unit normal to the LOS vector between the sensors. Kolitz[17] defines the reference plane to contain the origin of the inertial coordinate system and extends Varad's procedure for utilizing data from three sensors.

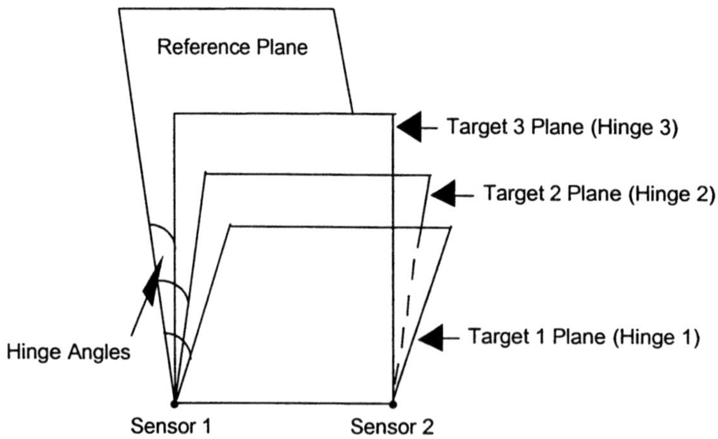

Figure 9.12: Varad hinge angle

Hinge angle association reduces the computational burden to a simple single parameter sort. The sorting parameter is the angle between the nominal reference plane and the plane containing the sensor and emitter target. The sets of scalar numbers, i.e., the direction cosines representing the degree of rotation of the reference plane into the sensor-emitter plane, are ordered into monotonic sequences, one sequence for each sensor. Ideally, in the absence of noise and when all sensors view all emitters, the sequences of the angles representing the emitters will be identical. Thus, there is a one-to-one correspondence between the two ordered sequences, resulting in association of the emitters. In the nonideal real-world application, the Varad algorithm matches up each emitter in the sequence of planes produced by one sensor, with an emitter having the closest hinge angle in the sequence produced by another sensor.

Once the hinge angles from the two sensors are associated, the range to the emitter is computed from the known angles between the LOS vectors to the emitter and the LOS vector between the sensors. Since the distance between the sensors is known, a triangle is defined with the emitter located at the third apex.

The range can be calculated using the law of sines as discussed in conjunction with Figure 9.4.

9.4 Passive computation of range using tracks from a single sensor site

Range to the target can also be estimated with track data from a single passive sensor that performs an appropriate maneuver.[21,22] In two dimensions, a Kalman filter using modified polar coordinates (MPC) is suitable for tracking non-maneuvering targets. These coordinates reduce problems associated with observability, range bias, and covariance ill-conditioning that are encountered with Cartesian coordinates. The MPC filter is extended to three dimensions by converting to modified spherical coordinates (MSC). The states in the Kalman tracking filter now include two angles, two angle rates, inverse time-to-go (equal to range rate divided by range), and inverse range. These states are transformable into Cartesian position and velocity.

The MSC filter can be applied to find the range to targets that are either non-maneuvering or maneuvering. If the target does not maneuver, the range state decouples from the other states in the tracking filter. If the target is maneuvering, then a batch estimation method (one that processes all of the observations simultaneously) is used to predict the future state of the target. Thus, maneuver detection must be an integral part of any successful range estimation algorithm in order to properly interpret the data from a single tracking sensor. A conventional approach to maneuver detection compares a chi-squared statistic based on the difference between the actual and expected measurement (also called the residual) with a threshold. If the chi-squared statistic is excessive (e.g., exceeds a confidence level of approximately 0.999), then a maneuver is declared present. A return to a nonmaneuver state occurs when the chi-squared statistic falls below a lower threshold. Other statistics are used to detect slow residual error accumulations.

9.5 Summary

Three data fusion techniques have been introduced for locating targets that emit energy. They are used with passive tracking systems or when it is anticipated that data otherwise available from active systems, including range information, will be unattainable. These techniques associate the data using either central or decentralized fusion architectures, with each having its own particular impact on data transmission and processing requirements.

In the first centralized fusion architecture, signals passively received by a surveillance radar and signals received through an auxiliary multibeam antenna are coherently processed. The resulting cross-correlation function expresses the likelihood that the signals received by the surveillance radar and multibeam antenna originated from the same source. This technique has a large impact on communications facilities because large-bandwidth raw signals need to be transmitted over potentially large distances. The impact on signal processing is equally large because of the range of time delays and Doppler shift that must be processed to include large search areas.

The second centralized fusion architecture combines azimuth direction angles or azimuth and elevation direction angles that are computed by each radar receiver. The major concern here is the elimination of ghost signals caused by noise or poor search geometry. Because processed data are transmitted to the central fusion processor, the communications channel bandwidth requirements are reduced as compared to those from the first architecture. The use of a maximum likelihood function allows the computation of an optimal solution for data association. The angle measurements are partitioned into two sets, one consisting of solutions corresponding to estimable emitter target positions and the other to spurious measurements. The final location of the emitter targets is found by converting the maximum likelihood formulation of the problem into a zero-one integer programming problem that is more easily solved. Here a zero is assigned to direction angle information from a radar that does not maximize the location of an emitter, and a one is assigned to information that does. The zero-one integer programming problem can be efficiently solved by applying constraints, such as using each radar's angle measurement data only once and eliminating variables that do not contribute to the maximization of the likelihood function. A suboptimal solution that significantly speeds up the data association process can also be found. This solution, which uses a relaxation algorithm, is particularly valuable when the number of emitters is large.

In the decentralized fusion architecture, still more processing is performed by individual radars located at distributed sites. High-confidence angle tracks of the emitter targets are formed at each site from the locally acquired sensor data using scan-to-scan target association or an auction algorithm. The tracks, along with unassociated data, are transmitted to a fusion center, where they are associated with the tracks and data sent from other sites. Two sensor-to-sensor track association methods were discussed: (1) a simple technique that eliminates sensor tracks already paired from further association, and (2) two global optimization techniques that allow all tracks from one sensor to be associated with all tracks from other sensors. The chi-squared value of the distance of closest approach of the tracks or the hinge angle is used to globally maximize the correct association of tracks and data received from the multiple sensor sites. Since most of the emitter location data have been reduced to tracks, the communications bandwidth required to transmit information to the fusion center is reduced even further. However, greater processing capability is required of the individual sensors. An alternate approach that allows range to be computed using angle tracks estimated by a single sensor was also discussed. This technique requires the tracking sensor to engage in a maneuver and to ascertain whether the tracked object has maneuvered or not.

References

1. D.E.N. Davies, "High data rate radars incorporating array signal processing and thinned arrays," *International Radar Conference Record*, Pub. 75CHO938-1 AES, IEEE, Piscataway, NJ (April 1975).
2. C.H. Knapp and C.C. Carter, "The generalized correlation method for estimation of time delay," *IEEE Trans. Acoust., Speech, Signal Processing*, A(4), 320-327 (August 1976).
3. J.S. Bendat and A.G. Piersol, *Engineering Applications of Correlation and Spectral Analysis*, John Wiley and Sons, New York, NY (1980).
4. H. Heidary, Personal communication.
5. Y. Bar-Shalom and T.E. Fortmann, *Tracking and Data Association*, Academic Press, New York, NY (1988).
6. P.R. Williams, "Multiple target estimation using multiple bearing-only sensors," *Proceedings of the 9th MIT/ONR Workshop on C3 Systems*, Report LIDS-R-1624, Massachusetts Institute of Technology, Laboratory for Information and Decision Systems, Cambridge, MA (December 1986).
7. E. Balas and M. Padberg, "Set partitioniong: a survey," *SIAM Rev.*, 18, 710-760 (October 1976).
8. J.F. Pierce and J.S. Lasky, "Improved combinatorial programming algorithms for a class of all zero-one integer programming problems," *Management Sci.*, 19, 528-543 (Jan. 1973).
9. R. Garfinkel and G. Nemhauser, *Integer Programming,* John Wiley and Sons, New York, NY (1972).
10. A.M. Frieze and J. Yadegar, "An algorithm for solving three-dimensional assignment problems with application to scheduling a teaching practice," *J. Op. Res. Soc.*, 32(11), 989-995 (1981).
11. P.R. Williams, personal communication.
12. K.R. Pattipati, S. Deb, Y. Bar-Shalom, and R.B. Washburn, Jr., "A new relaxation algorithm and passive sensor data association," *IEEE Trans. Automat. Control*, 37, 198-213 (Feb. 1992).
13. D.B. Reid, "An algorithm for tracking multiple targets," *IEEE Trans. Automat. Control*, AC-24, 423-432 (Dec. 1979).
14. E. Lawler and E. Wood, "Branch-and-bound methods: A survey," *J. Op. Res.*, 14, 699-719 (1966).
15. R. Babuska, *Fuzzy Modeling for Control*, Kluwer Academic Publishers, Boston, MA (1998).
16. R. Varad, "Scalar correlation algorithm: multi-target, multiple-sensor data fusion," *Proc. of the 1st National Symposium on Sensor Fusion*, GACIAC PR-88-01, 433-441 (April 1988).
17. S.E. Kolitz, "Passive sensor data fusion," *Signal and Data Processing of Small Targets, Proc. SPIE* 1481, 329-340 (1991).
18. J. Munkres, "Algorithms for the assignment and transportation problem," *J. SIAM*, 5, 32-38 (March 1957).

19. F. Bourgeois and J.C. Lassalle, "An extension of the Munkres algorithm for the assignment problem to rectangular matrices," *Comm. Association for Computing Machinery*, 802-804 (1971).
20. O.E. Drummond, D.A. Castanon, and M.S. Bellovin, "Comparison of 2-D assignment algorithms for sparse, rectangular, floating point, and cost matrices," *J. SDI Panels Tracking,* Institute for Defense Analyses, Alexandria, VA, Issue No. 4/1990, 4-81 to 4-97 (Dec. 15, 1990).
21. D.V. Stallard, "Angle-only tracking filter in modified spherical coordinates," *J. Guidance*, 14(3), 694-696 (May-June 1991).
22. R.R. Allen and S.S. Blackman, "Implementation of an angle-only tracking filter," *Signal and Data Processing of Small Targets*, *Proc. SPIE* 1481, 292-303 (1991).

Appendix A

Planck Radiation Law and Radiative Transfer

A.1 Planck's radiation law

Blackbody objects (i.e., perfect emitters of energy) above absolute zero emit energy E per unit volume and per unit frequency at all wavelengths according to the Planck radiation law

$$E = \frac{8\pi hf^3}{c^3} \frac{1}{\exp\left(\frac{hf}{k_B T}\right) - 1} \frac{\text{J}}{\text{m}^3\text{Hz}} \tag{A-1}$$

where

h = Planck's constant = 6.6256×10^{-34} J-s,

k_B = Boltzmann's constant = 1.380662×10^{-23} J/K,

c = speed of light = 3×10^8 m/s,

T = physical temperature of the emitting object in degrees K,

f = frequency at which the energy is measured in Hz.

Upon expanding the exponential term in the denominator, Eq. (A-1) may be rewritten as

$$E = \frac{8\pi hf^3}{c^3} \frac{1}{\frac{hf}{k_B T} + \left(\frac{hf}{k_B T}\right)^2 + \ldots} \frac{\text{J}}{\text{m}^3\text{Hz}}. \tag{A-2}$$

For frequencies f less than $k_B T/h$ ($\approx 6 \times 10^{12}$ Hz at 300 K), only the linear term in temperature is retained and Planck's radiation law reduces to the Rayleigh-Jeans law given by

$$E = \frac{8\pi f^2 k_B T}{c^3} \frac{\text{J}}{\text{m}^3\text{Hz}}. \tag{A-3}$$

In the Rayleigh-Jeans approximation, temperature is directly proportional to the energy of the radiating object, making calibration of a radiometer simpler. With perfect emitters or blackbodies, the physical temperature of the object T is equal to the brightness temperature T_B that is detected by a radiometer. However, the surfaces of real objects do not normally radiate as blackbodies (i.e., they are not 100 percent efficient in emitting the energy predicted by the Planck radiation law). To account for this nonideal emission, a multiplicative emissivity factor is added to represent the amount of energy that is radiated by the object, now referred to as a graybody. The emissivity is equal to the ratio of T_B to T.

When microwave radiometers are used in space applications, the first three terms of the exponential series (up to and including the second-order term containing $[hf/k_BT]^2$) in Eq. (A-2) are retained, because the background temperature of space is small compared to the background temperatures on Earth. Including the quadratic term in temperature minimizes the error that would otherwise occur when converting the measured energy into atmospheric temperature profiles used in weather forecasting. The magnitude of the error introduced when the quadratic term is neglected is shown in Table A.1 as a function of frequency.

Table A.1: Effect of quadratic correction term on emitted energy calculated from Planck radiation law ($T = 300$ K)

f (GHz)	hf/k_BT	$(hf/k_BT)^2$	% change in E
2	0.0003199	1.0233×10^{-07}	0.03198812
6	0.0009598	9.2122×10^{-07}	0.09598041
22	0.0035192	1.2385×10^{-05}	0.35192657
60	0.0095977	9.2116×10^{-05}	0.95977161
118	0.0188755	0.00035628	1.88752616
183	0.0292730	0.00085691	2.92730503
320	0.0511878	0.00262019	5.11877830

Because of emission from molecules not at absolute zero, the atmosphere emits energy that is detected by passive sensors that directly or indirectly (such as by reflection of energy from surfaces whose emissivity is not unity) view the atmosphere. The atmospheric emission modifies and may prevent the detection of ground-based and space-based objects of interest by masking the energy emitted by low-temperature or low-emissivity objects. However, radiometers used in weather forecasting applications operate at atmospheric absorption bands in order to measure the quantity of atmospheric constituents such as oxygen and water vapor. In both cases, radiative transfer theory is used to calculate the effects of the atmosphere on the energy measured by the radiometer.

A.2 Radiative transfer theory

Radiative transfer theory describes the contribution of cosmic, galactic, atmospheric, and ground-based emission sources to the passive signature of objects in a sensor's field of view.[1,2] In Figure A.1, a radiometer is shown flying in a missile or gun-fired round at a height *h* above the ground and is pointed toward the ground. If the application were weather forecasting, the radiometer would be located in a satellite.

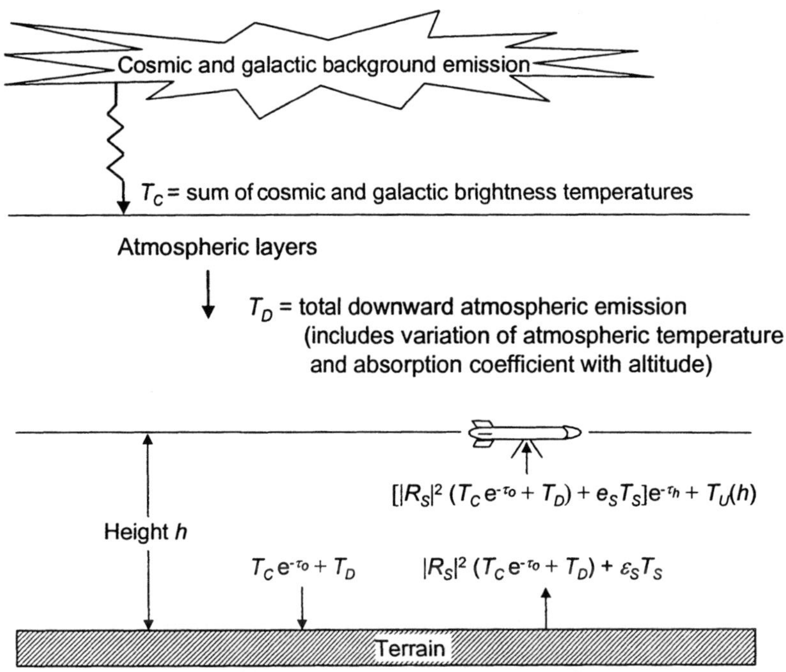

Figure A.1: Radiative transfer in an Earth-looking radiometer sensor

The quantity T_C represents the sum of the cosmic brightness temperature and the galactic brightness temperature. The cosmic temperature is independent of frequency and zenith angle and is equal to 2.735 K. Its origin is attributed to the background radiation produced when the universe was originally formed. The galactic temperature is due to radiation from the Milky Way galaxy and is a function of the viewing direction and frequency. Above about 10 GHz, the galactic contribution may be neglected in comparison with the downward emission from the atmosphere.

The variable T_D represents the total downward atmospheric emission, including the variation of temperature $T(z)$ and absorption coefficient $K_\alpha(z)$ with height, and is equal to

$$T_D = \int_0^{\infty} \{T(z)\exp[-\sec\theta \int_0^{z} K_{\alpha}(z')\,dz']K_{\alpha}(z)\}\,dz\sec\theta\,, \tag{A-4}$$

where

$K_\alpha(z)$ = absorption coefficient of the atmosphere at an altitude z and

θ = incidence angle with respect to nadir as shown in Figure A.2.

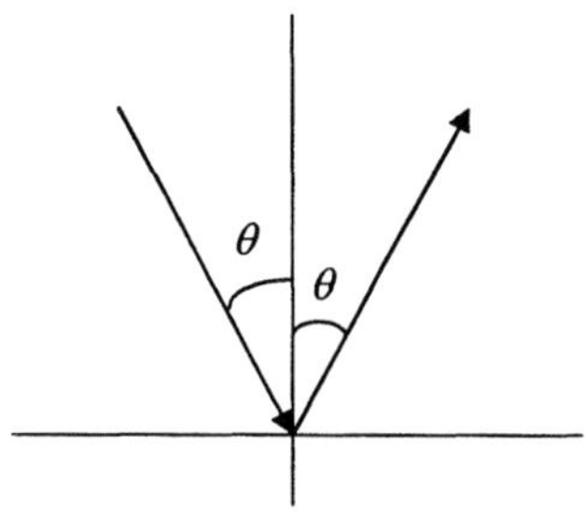

Figure A.2: Definition of incidence angle θ

The dependence of the absorption coefficient on altitude accounts for the energy emitted by the atmosphere through its constituent molecules such as water vapor, oxygen, ozone, carbon dioxide, and nitrous oxide. Since emission occurs throughout the entire atmospheric height profile, the integration limits for T_D are from 0 to infinity.

The quantity T_U is the upward atmospheric emission in the region from the ground to the height h at which the sensor is located. It is given by

$$T_U = \int_0^{h} \{T(z)\exp[-\sec\theta \int_0^{z} K_{\alpha}(z')\,dz']K_{\alpha}(z)\}\,dz\sec\theta\,. \tag{A-5}$$

The quantity τ_o is the total one-way opacity (integrated attenuation) through the atmosphere. When $\theta < 70$ deg and the atmosphere is spherically stratified, τ_o may be expressed as

$$\tau_o = \int_0^{\infty} K_{\alpha}(z)\,dz\sec\theta\,. \tag{A-6}$$

The quantity τ_h represents the one-way opacity from ground to height h equal to

$$\tau_h = \int_0^{h} K_{\alpha}(z)\,dz\sec\theta\,. \tag{A-7}$$

The variable R_S is the Fresnel reflection coefficient at the atmosphere-ground interface. The square of its magnitude is the reflectivity, which is expressed as

$$|R_S|^2 = 1 - \varepsilon_S, \tag{A-8}$$

where ε_S is the emissivity of the Earth's surface in the field of view of the sensor. The emissivity is a function of the operating frequency, polarization, and viewing angle of the sensor. Perfect emitters of energy, i.e., blackbody radiators, have an emissivity of one. Perfect conductors, such as shiny metal objects, have an emissivity of zero. Most objects are graybodies and have emissivities inbetween these limits.

The first term in the energy detected by the radiometer (as shown entering the radiometer receiver in Figure A.1) is produced by the cosmic and galactic brightness temperatures. Both are attenuated by τ_o and, along with the total downward atmospheric emission, are reflected from the ground upward toward the radiometer. In addition to the reflected energy, the brightness temperature T_B emitted by the ground (or a target if one is present within the footprint of the radiometer) contributes to the energy received by the radiometer. It is equal to the product $\varepsilon_S T_S$, where T_S is the absolute temperature of the ground surface.

The opacity of the intervening atmosphere τ_h between the ground and the radiometer at height h attenuates the energy sources described above before they reach the radiometer. To these is added the upward emission $T_U(h)$ due to atmospheric absorption phenomena that exist between the ground and height h. The total energy E received by the radiometer at a height h above the ground surface is found by combining the above energy sources as

$$E = [|R_S|^2 (T_C e^{-\tau_o} + T_D) + e_S T_S] e^{-\tau_h} + T_U(h) \,. \tag{A-9}$$

Two simplifications to the general radiative transfer equation of (A-9) can be made when the radiometer is deployed at low altitudes. First the cosmic, galactic, and downwelling atmospheric emission terms can be combined into one term called the sky radiometric temperature denoted by T_{sky}. The temperature T_{sky} is still a function of atmospheric water content, cloud cover, and radiometer operating frequency. A summary of the downwelling atmospheric temperature and atmospheric attenuation is given in Table A-2 for a zenith-looking radiometer under clear air and 11 types of cloud conditions at *S*-band, *X*-band, and *Ka*-band frequencies.[3] As discussed in Chapter 2, both the downwelling atmospheric temperature and atmospheric attenuation increase with increasing frequency and increasing water content of the clouds. The cosmic and galactic temperatures are not included in the atmospheric temperature shown in the table. The second simplification that occurs when a radiometer is deployed at low altitudes (e.g., as part of a suite of sensors in a surface-to-surface missile) is made possible by neglecting the small upwelling atmospheric contribution.

Table A.2: Downwelling atmospheric temperature T_D and atmospheric attenuation A for a zenith-looking radiometer

(S.D. Slobin, *Microwave Noise Temperature and Attenuation of Clouds at Frequencies Below 50 GHz*, JPL Publication 81-46, Jet Propulsion Laboratory, Pasadena, CA [July 1, 1981])

Case	Lower Cloud				Upper Cloud					S-Band (2.3 GHz) Zenith		X-Band (8.5 GHz) Zenith		Ka-Band (32 GHz) Zenith	
	Density g/m³	Base km	Top km	Thickness km	Density g/m³	Base km	Top km	Thickness km	Remarks	T_D (K)	A (dB)	T_D (K)	A (dB)	T_D (K)	A (dB)
1	–	–	–	–	–	–	–	–	Clear Air	2.15	0.035	2.78	0.045	14.29	0.228
2	0.2	1.0	1.2	0.2	–	–	–	–	Light, Thin Clouds	2.16	0.036	2.90	0.047	15.92	0.255
3	–	–	–	–	0.2	3.0	3.2	0.2		2.16	0.036	2.94	0.048	16.51	0.266
4	0.5	1.0	1.5	0.5	–	–	–	–		2.20	0.036	3.55	0.057	24.56	0.397
5	–	–	–	–	0.5	3.0	3.5	0.5		2.22	0.037	3.83	0.062	28.14	0.468
6	0.5	1.0	2.0	1.0	–	–	–	–	Medium Clouds	2.27	0.037	4.38	0.070	35.22	0.581
7	–	–	–	–	0.5	3.0	4.0	1.0		2.31	0.038	4.96	0.081	42.25	0.731
8	0.5	1.0	2.0	1.0	0.5	3.0	4.0	1.0		2.43	0.040	6.55	0.105	61.00	1.083
9	0.7	1.0	2.0	1.0	0.7	3.0	4.0	1.0		2.54	0.042	8.04	0.130	77.16	1.425
10	1.0	1.0	2.0	1.0	1.0	3.0	4.0	1.0	Heavy Clouds	2.70	0.044	10.27	0.166	99.05	1.939
11	1.0	1.0	2.5	1.5	1.0	3.5	5.0	1.5		3.06	0.050	14.89	0.245	137.50	3.060
12	1.0	1.0	3.0	2.0	1.0	4.0	6.0	2.0	Very Heavy Clouds	3.47	0.057	20.20	0.340	171.38	4.407

Cases 2-12 are clear air and clouds combined.
Antenna located at sea level, heights are measured from ground level.
Cosmic and galactic brightness temperatures and ground contributions are not included in the downwelling temperature T_D.
Attenuation A is measured along a vertical path from ground to 30-km altitude.

References

1. F.T. Ulaby, R.K. Moore, and A.K. Fung, *Microwave Remote Sensing: Active and Passive, Vol. I, Microwave Remote Sensing Fundamentals and Radiometry*, Artech House, Norwood, MA (1981).
2. G.M. Hidy, W.F. Hall, W.N. Hardy, W.W. Ho, A.C. Jones, A.W. Love, M.J. Van Melle, H.H. Wang, and A.E. Wheeler, *Development of a Satellite Microwave Radiometer to Sense the Surface Temperaure of the World Oceans*, NASA Contractor Rpt. CR-1960, National Aeronautics ansd Space Administration, Washington, DC (February 1972).
3. S.D. Slobin, *Microwave Noise Temperature and Attenuation of Clouds at Frequencies Below 50 GHz*, JPL Publication 81-46, Jet Propulsion Laboratory, Pasadena, CA (July 1, 1981).

Appendix B

Voting Fusion with Nested Confidence Levels

The key to applying Eq. (7-9) in Chapter 7 is the creation of nonnested confidence levels for each sensor as was illustrated in Figure 7.3. These allow a unique value to be selected for the inherent sensor detection probability when different signal-to-interference ratios are postulated and implemented at each confidence level. In fact, the ability to specify and then implement unique detection probabilities for each confidence level makes this voting fusion technique practical.

However, a Venn diagram such as the one in Figure B.1 with nested confidence levels implies that the detection probabilities at each confidence level are not independent.

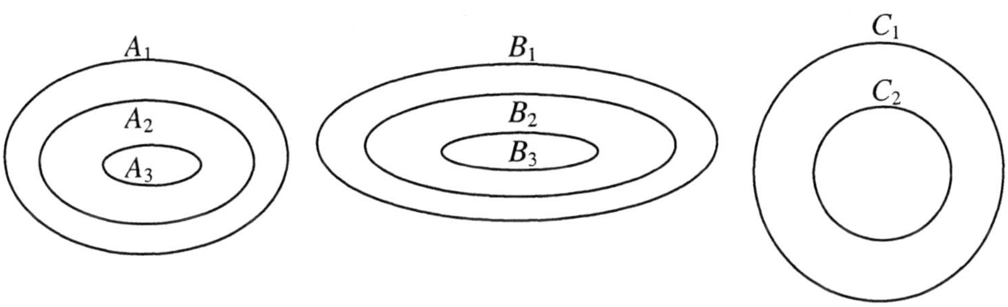

Figure B.1: Nested sensor confidence levels

Here, the confidence levels A_1, A_2, and A_3 of Sensor A and the confidence levels in the other sensors are not independent of each other. Confidence level A_3 is a subset of level A_2, which is a subset of level A_1. Discriminants other than signal-to-interference ratio are used in this case to differentiate the confidence levels. For example, target-like features that are present in the signal can be exploited by algorithms to increase the confidence that the signal belongs to a bonafide target. This model is more restrictive and may not depict the way the sensors are actually operating in a particular application.

A different Boolean expression is also needed to compute the detection probability of the three-sensor suite when nested confidence levels are postulated. Since the confidence levels for each sensor are not independent, the simplifying assumptions of Eqs. (7-4) and (7-5) no longer apply. The Boolean equation for the system detection probability with nested confidence levels takes the form

$$\text{System } P_d = P_d\{A_1 B_1 C_1 \text{ or } A_2 C_2 \text{ or } B_2 C_2 \text{ or } A_3 B_3) \tag{B-1}$$

or

$$\begin{aligned}\text{System } P_d = {} & P_d\{A_1 B_1 C_1\} + P_d\{A_2 C_2\} + P_d\{B_2 C_2\} + P_d\{A_3 B_3\} \\ & - P_d\{A_2 B_1 C_2\} - P_d\{A_1 B_2 C_2\} - P_d\{A_3 B_3 C_1\}.\end{aligned} \tag{B-2}$$

If the sensors respond to independent signature-generation phenomena such that the likelihood of detection by one sensor is independent of that of another, then

$$\begin{aligned}\text{System } P_d = {} & P_d\{A_1\}\, P_d\{B_1\}\, P_d\{C_1\} + P_d\{A_2\}\, P_d\{C_2\} + P_d\{B_2\}\, P_d\{C_2\} \\ & + P_d\{A_3\}\, P_d\{B_3\} - P_d\{A_2\}\, P_d\{B_1\}\, P_d\{C_2\} \\ & - P_d\{A_1\}\, P_d\{B_2\}\, P_d\{C_2\} - P_d\{A_3\}\, P_d\{B_3\}\, P_d\{C_1\}.\end{aligned} \tag{B-3}$$

The difference terms represent areas of overlap that are accounted for more than once in the sum terms.

The false alarm probability of the three-sensor system is also in the form of (B-3) with P_d replaced by P_{fa}. Thus, with nested confidence levels, the system false alarm probability is

$$\begin{aligned}\text{System } P_{fa} = {} & P_{fa}\{A_1\}\, P_{fa}\{B_1\}\, P_{fa}\{C_1\} + P_{fa}\{A_2\}\, P_{fa}\{C_2\} + P_{fa}\{B_2\}\, P_{fa}\{C_2\} \\ & + P_{fa}\{A_3\}\, P_{fa}\{B_3\} - P_{fa}\{A_2\}\, P_{fa}\{B_1\}\, P_{fa}\{C_2\} \\ & - P_{fa}\{A_1\}\, P_{fa}\{B_2\}\, P_{fa}\{C_2\} - P_{fa}\{A_3\}\, P_{fa}\{B_3\}\, P_{fa}\{C_1\}.\end{aligned} \tag{B-4}$$

Index

A

B

C

D

E

M

Q

R

S

T

Lawrence A. Klein received the B.E.E. degree from the City College of New York in 1963, the M.S. degree in electrical engineering from the University of Rochester in 1966, and the Ph.D. degree in electrical engineering from New York University in 1973 along with the Founders Day Award for outstanding academic achievement.

Dr. Klein is a consultant, specializing in data and sensor fusion applications to aerospace and transportation management and the design of efficient multimodal ground transportation systems.

He formerly worked at Hughes Aircraft (now Raytheon) in Fullerton, CA, where he managed Intelligent Transportation System programs. While at Hughes, Dr. Klein also developed fusion concepts for tactical and reconnaissance systems that incorporated multiple sensors and communications media. He worked as systems manager at Aerojet, where he was responsible for the conceptual design and program development of satellite- and ground-based multiple sensor systems. As program manager of smart munitions programs, he developed sensor fusion hardware and signal processing algorithms for systems using millimeter-wave and infrared radars and radiometer sensors. He also served as program manager of manufacturing methods and technology programs that lowered costs of GaAs monolithic microwave integrated circuit components. Before joining Aerojet, Dr. Klein was employed at Honeywell where he developed polarimetric radar and passive millimeter-wave monopulse tracking systems. His prior work at NASA Langley Research Center explored remote sensing of terrain and weather from both phenomenology and sensor development aspects.

Dr. Klein regularly teaches short courses in advanced sensors and data fusion, including those for the SPIE, Johns Hopkins University, and ITS-America. He is a member of Eta Kappa Nu and has been a reviewer for the IEEE *Transactions on Antennas and Propagation* and *Geoscience and Remote Sensing*. He has also served as Chairman of the Minneapolis–St. Paul Twin Cities Chapter of the Communications, Speech, and Signal Processing Society and Education Chairman of the Twin Cities Section. He is a licensed professional engineer (New York).